INTRODUCTION TO

Quantum Field Theory
and the Standard Model

Recommended Titles in Related Topics

Lectures of Sidney Coleman on Quantum Field Theory
Foreword by David Kaiser
edited by Bryan Gin-ge Chen, David Derbes, David Griffiths,
Brian Hill, Richard Sohn and Yuan-Sen Ting
ISBN: 978-981-4632-53-9
ISBN: 978-981-4635-50-9 (pbk)

Lectures on Quantum Field Theory
Second Edition
by Ashok Das
ISBN: 978-981-122-086-9
ISBN: 978-981-122-216-0 (pbk)

Foundations of Quantum Field Theory
by Klaus D Rothe
ISBN: 978-981-122-192-7
ISBN: 978-981-122-300-6 (pbk)

Mathematical Foundations of Quantum Field Theory
by Albert Schwarz
ISBN: 978-981-3278-63-9

———— INTRODUCTION TO ————
Quantum Field Theory
and the Standard Model

Wolfgang Hollik

Max Planck Institute for Physics, Germany

W꜀ World Scientific

NEW JERSEY · LONDON · SINGAPORE · BEIJING · SHANGHAI · HONG KONG · TAIPEI · CHENNAI · TOKYO

Published by

World Scientific Publishing Co. Pte. Ltd.
5 Toh Tuck Link, Singapore 596224
USA office: 27 Warren Street, Suite 401-402, Hackensack, NJ 07601
UK office: 57 Shelton Street, Covent Garden, London WC2H 9HE

Library of Congress Cataloging-in-Publication Data
Names: Hollik, W. (Wolfgang), 1951–   author.
Title: Introduction to quantum field theory and the standard model /
    Wolfgang Hollik, Max Planck Institute for Physics, Germany.
Description: New Jersey : World Scientific, [2022] |
    Includes bibliographical references and index.
Identifiers: LCCN 2021050559 (print) | LCCN 2021050560 (ebook) |
    ISBN 9789811242175 (hardcover) | ISBN 9789811242182 (ebook) |
    ISBN 9789811242199 (ebook other)
Subjects: LCSH: Quantum field theory.
Classification: LCC QC174.45 .H647 2022 (print) | LCC QC174.45 (ebook) |
    DDC 530.14/3--dc23/eng/20211201
LC record available at https://lccn.loc.gov/2021050559
LC ebook record available at https://lccn.loc.gov/2021050560

British Library Cataloguing-in-Publication Data
A catalogue record for this book is available from the British Library.

For any available supplementary material, please visit
https://www.worldscientific.com/worldscibooks/10.1142/12415#t=suppl

Typeset by Stallion Press
Email: enquiries@stallionpress.com

Preface

These lecture notes are based on material presented during courses at the Technical University Munich (TUM) to physics students with interest in particle physics and its theoretical foundation. This material has been elaborated and expanded, to provide both the conceptual ideas regarding the Standard Model of particle physics and some technical details on the formulation within the context of gauge theories. The scope ranges from relativistic quantum mechanics to an introduction to quantum field theory with quantum electrodynamics as an example of a successful theory of a fundamental interaction, and is further extended towards quantum chromodynamics for the strong interaction and to the electroweak Standard Model for the unified electromagnetic and weak interactions, which could celebrate an enormous success by the 2012 discovery of a new spin-0 particle showing the properties of the predicted Higgs boson.

Quantum field theory is the primary theoretical tool for the description of the microscopic dynamics of the strong and electroweak interactions. It is thus necessary to provide an introduction to the basic concepts of relativistic quantum field theory, perturbation theory, Feynman graphs, before advancing to Abelian and non-Abelian gauge theories and their application to the fundamental forces and the resulting phenomenological implications with their experimental tests. The success of perturbative QCD at high energies goes with the basic features of asymptotic freedom and the parton model, displayed in the chapter on QCD. Phenomenology of W and Z bosons as well as Higgs bosons is part of the electroweak chapter including recent experimental results, precision tests and current status of the Standard Model.

The style is elementary and pedagogical. The text is at a level appropriate for students who had already a course in quantum mechanics and are familiar with classical electrodynamics and the basics of special relativity. Special relativity in the first chapter is a compact recapitulation serving as a summary and for setting the language and notations. The formalism of quantum field theory is kept at a minimum. Especially the Feynman rules are not derived in a systematic way; instead, the method is elucidated on the basis of examples to convey a more intuitive understanding that is thought to be more helpful for an early overview rather than proofs and advanced technical effort. Explicit calculations are performed for selected examples so that the reader becomes acquainted with practical calculations in particular for scattering amplidudes, cross sections and decay rates at lowest order. Extra material on specific theoretical issues is included in various places for the interested student; it can be skipped during a first reading.

The lecture notes provide a compact introduction, convenient for students dealing with particle physics and related areas to get first information before specializing towards experimental or theoretical topics. They may also serve as a basis to study elaborate textbooks, like Michael Peskin and Daniel Schroeder's *An Introduction to Quantum Field Theory* (Westview Press, 1995), Matthew Schwartz's *Quantum Field Theory and the Standard Model* (Cambridge University Press, 2014), or Steven Weinberg's *The Quantum Theory of Fields* (Cambridge University Press, 1995).

It is my special concern to thank my colleagues from the Physics Department at TUM, Peter Fierlinger, Lothar Oberauer, Stephan Paul, and Stefan Schönert, for the pleasant cooperation during various lecture courses covering both experimental and theoretical aspects, over many years of teaching.

Wolfgang Hollik

Contents

Chapter 1

Special Relativity

It is a assumed that the reader is already familiar with special relativity, e.g. at the level of a first course on electrodynamics. Thus, this chapter serves essentially as a summary of the relevant theoretical ingredients, to introduce notations and conventions, and to provide the classical basis for advancing to quantum theory in the subseqent chapters.

The Special Theory of Relativity is based on the following two fundamental principles.

(1) *Principle of relativity*
The laws of nature are in all inertial frames of the same form.

(2) *Invariance of light propagation*
The propagation of light is independent of the inertial frame. For a pointlike source at a space point $\vec{x}_0$ and time t_0 in a given frame K the outgoing lightfronts are spherical surfaces in all inertial frames, and propagate with c, the constant velocity of light,

$$\underbrace{c^2(t - t_0)^2 - (\vec{x} - \vec{x}_0)^2}_{\text{inertial frame } K} = 0 = \underbrace{c^2(t' - t_0')^2 - (\vec{x}\,' - \vec{x}_0')^2}_{\text{inertial frame } K'}. \qquad (1.1)$$

Principle (2) is not valid for Galilei transformations between different frames. Thus, they have to be replaced by Lorentz transformations.

1.1 Notations and Conventions

In a given inertial frame a space-time event is determined by a set of coordinates specifying the point $\vec{x}$ in space and t in time. This information can be summarized by a four-component quantity (x^μ) according to

$$(x^\mu) = (x^0, x^1, x^2, x^3) \equiv (x^0, \vec{x}) \tag{1.2}$$

with

$$x^0 = ct, \quad \vec{x} = (x^1, x^2, x^3). \tag{1.3}$$

The (x^μ) form a 4-dimensional linear space, with elements denoted as 4-vectors and labeled by the symbol $x \equiv (x^\mu)$ as a short-hand notation. The quantities x^μ are called *contravariant components*.

Convention: greek indices $\mu, \nu, \ldots = 0, 1, 2, 3$
                     latin indices $k, l, \ldots = 1, 2, 3$.

Metric. For the space of 4-vectors a scalar product is defined as a symmetric bilinear form as follows,

$$(x, y) \rightarrow x \cdot y = x^0 y^0 - \vec{x} \cdot \vec{y} = g_{\mu\nu} x^\mu y^\nu \tag{1.4}$$

with the metric tensor

$$(g_{\mu\nu}) = \begin{pmatrix} 1 & 0 & 0 & 0 \\ 0 & -1 & 0 & 0 \\ 0 & 0 & -1 & 0 \\ 0 & 0 & 0 & -1 \end{pmatrix}. \tag{1.5}$$

The square of a 4-vector is thus given by

$$x^2 = x \cdot x = g_{\mu\nu} x^\mu x^\nu = (x^0)^2 - \vec{x}^2. \tag{1.6}$$

The 4-dimensional space of the (x^μ) with this metric is denoted as *Minkowski space*, the metric (1.4) as *Minkowski metric*.

Convention. In Eq. (1.4) and also in the following, the Einstein convention is used that summation is performed over upper and lower indices when they are equal.

Covariant components. The covariant components of a 4-vector are defined by means of the metric tensor as follows,

$$x_\mu = g_{\mu\nu} x^\nu, \quad \text{yielding } x_0 = x^0, \ x_k = -x^k. \tag{1.7}$$

With this notation the square of a 4-vector can be written in the following way,

$$x^2 = x_\mu x^\mu = x^\mu x_\mu, \tag{1.8}$$

as well as the scalar product, respectively,

$$x \cdot y = x_\mu y^\mu = x^\mu y_\mu. \tag{1.9}$$

1.2 Lorentz Transformations

A Lorentz transformation from an inertial frame K into another inertial frame K' is described by a linear transformation of the components x^μ in K to x'^μ in K' with the help of a matrix $\Lambda = (\Lambda^\mu_{\ \nu})$,

$$\boxed{x'^\mu = \Lambda^\mu_{\ \nu}\, x^\nu}. \tag{1.10}$$

The invariance of light propagation (1.1) is equivalent to the invariance of the scalar product under Lorentz transformations, which can be fomulated as a condition

$$g_{\rho\sigma}\, x^\rho x^\sigma = g_{\mu\nu}\, x'^\mu x'^\nu = g_{\mu\nu}\, \Lambda^\mu_{\ \rho} \Lambda^\nu_{\ \sigma}\, x^\rho x^\sigma \tag{1.11}$$

to be fulfilled for all x^μ. Hence, Eq. (1.11) is a condition for the entries of Λ,

$$g_{\mu\nu}\, \Lambda^\mu_{\ \rho} \Lambda^\nu_{\ \sigma} = g_{\rho\sigma}, \tag{1.12}$$

or expressed in a compact way using the matrix $g = (g_{\mu\nu})$ as follows,

$$\Lambda^T g\, \Lambda = g. \tag{1.13}$$

Applying the rules for determinants,

$$\det(\Lambda^T g\, \Lambda) = \det(\Lambda^T)\det(g)\det(\Lambda) = \det(g), \tag{1.14}$$

it immediately follows for the determinant of a Lorentz transformation that

$$\det(\Lambda) = \pm 1. \tag{1.15}$$

As a consequence, the 4-dimensional volume element is invariant under Lorentz transformations,

$$\mathrm{d}^4 x = \mathrm{d}x^0\, \mathrm{d}^3 x \quad \rightarrow \quad \mathrm{d}^4 x' = |\det(\Lambda)|\, \mathrm{d}^4 x = \mathrm{d}^4 x, \tag{1.16}$$

because Λ is the Jacobian matrix of the transformation of variables $x^\mu \rightarrow x'^\mu$.

1.2.1 *Examples of Lorentz tranformations*

(i) *Space inversion* $x^k \rightarrow -x^k,\ x^0 \rightarrow x^0$
   $\det(\Lambda) = -1$.

(ii) *Time reversal* $x^0 \rightarrow -x^0,\ x^k \rightarrow x^k$
   $\det(\Lambda) = -1$.

(iii) *Rotation*

$$\Lambda = \begin{pmatrix} 1 & 0 & 0 & 0 \\ 0 & & & \\ 0 & & \mathbf{R} & \\ 0 & & & \end{pmatrix} \qquad \text{with } \mathbf{R} = \text{rotation matrix} \qquad (1.17)$$

$\det(\Lambda) = \det(\mathbf{R}) = +1$.

(iv) *Boost*   Frame K' moving in frame K with constant velocity $\vec{v}$.
$\det(\Lambda) = +1$.

Specific: Boost in x^1-direction, K' moving in K with v in positive x^1-direction

$$\Lambda = \begin{pmatrix} \gamma & -\beta\gamma & 0 & 0 \\ -\beta\gamma & \gamma & 0 & 0 \\ 0 & 0 & 1 & 0 \\ 0 & 0 & 0 & 1 \end{pmatrix} \qquad \text{with } \beta = \frac{v}{c}, \ \gamma = \frac{1}{\sqrt{1-\beta^2}} \qquad (1.18)$$

With the help of the *rapidity* ϕ the boost matrix can be written as follows,

$$\Lambda = \begin{pmatrix} \cosh\phi & -\sinh\phi & 0 & 0 \\ -\sinh\phi & \cosh\phi & 0 & 0 \\ 0 & 0 & 1 & 0 \\ 0 & 0 & 0 & 1 \end{pmatrix} \qquad \text{with } \tanh\phi = \frac{v}{c}. \qquad (1.19)$$

On the geometrical meaning of the components. Consider a 4-dimensional real vector space $\mathbf{V}$, with basis vectors $\underline{e}_\mu$ ($\mu = 0, 1, 2, 3$). Each vector $\underline{x} \in \mathbf{V}$ can be expanded in terms of the basis vectors, $\underline{x} = x^\mu \underline{e}_\mu$, where the coefficients $x^\mu \in \mathbb{R}$ are the *contravariant components* of the vector $\underline{x}$. When a basis transformation $\underline{e}_\mu \to \underline{e}'_\mu$ is performed according to

$$\underline{e}_\nu = \Lambda^\mu{}_\nu \, \underline{e}'_\mu \quad (\nu = 0, 1, 2, 3) \qquad (1.20)$$

one obtains for the components

$$x'^\mu = \Lambda^\mu{}_\nu \, x^\nu, \qquad (1.21)$$

because

$$\underline{x} = x^\nu \underline{e}_\nu = \underbrace{x^\nu \Lambda^\mu{}_\nu}_{x'^\mu} \, \underline{e}'_\mu = x'^\mu \underline{e}'_\mu.$$

The contravariant components are thus transformed inversely to the basis. In case of a metric vector space with a scalar product $\underline{x}\cdot\underline{y} \in \mathbb{R}$ one can define the scalar product for the basis vectors $\underline{e}_\mu$ by

$$\underline{e}_\mu\cdot\underline{e}_\nu = g_{\mu\nu} = g_{\nu\mu}. \tag{1.22}$$

Then $\underline{x}\cdot\underline{y}$ is determined for all $\underline{x}$ and $\underline{y} \in \mathbf{V}$,

$$\underline{x}\cdot\underline{y} = (x^\mu\underline{e}_\mu)\cdot(y^\nu\underline{e}_\nu) = x^\mu y^\nu g_{\mu\nu}. \tag{1.23}$$

In particular one obtains

$$\underline{x}\cdot\underline{e}_\mu = x^\nu\underline{e}_\nu\cdot\underline{e}_\mu = x^\nu g_{\nu\mu} = g_{\mu\nu}x^\nu = x_\mu, \tag{1.24}$$

which means that the covariant components x_μ are the projections of $\underline{x}$ on the basis vectors. For a basis transformation $\underline{e}_\mu \to \underline{e}'_\mu$ given in Eq. (1.20) the relation

$$x_\nu = \underline{x}\cdot\underline{e}_\nu = \underline{x}\cdot(\Lambda^\mu_{\ \nu}\,\underline{e}'_\mu) = \Lambda^\mu_{\ \nu}(\underline{x}\cdot\underline{e}'_\mu) = \Lambda^\mu_{\ \nu}\,x'_\mu \tag{1.25}$$

determines the transformation

$$\boxed{x_\nu = x'_\mu\,\Lambda^\mu_{\ \nu}}. \tag{1.26}$$

Accordingly, the covariant components are transformed in the same way as the basis vectors (1.20) and inversely to the contravariant components.

In Minkowski space, the basis vectors define the reference frame. We will always use the component notation for vectors.

1.2.2 *General 4-vector*

A 4-vector in general is a 4-component quantity a^μ that transforms under a Lorentz transformation from K to K' like the components x^μ, i.e. with the matrix $(\Lambda^\mu_{\ \nu})$ as follows,

$$a'^\mu = \Lambda^\mu_{\ \nu}\,a^\nu. \tag{1.27}$$

Accordingly, a^μ are denoted as the contravariant components. Moreover, one has

$$a_\mu = g_{\mu\nu}a^\nu, \quad a^2 = a_\mu a^\mu = (a^0)^2 - \vec{a}^2, \quad \vec{a} = (a^1, a^2, a^3) \tag{1.28}$$

for the covariant components and the invariante square.

1.2.3 *Tensor*

A quantity with two indices is denoted as a (contravariant) tensor of rank two when under a Lorentz transformation (1.10) each index transforms with the matrix Λ,

$$T'^{\mu\nu} = \Lambda^{\mu}_{\ \rho}\Lambda^{\mu}_{\ \sigma}\, T^{\rho\sigma}. \qquad (1.29)$$

The concept can be easily extended to tensors of higher rank, $T^{\mu_1\mu_2\cdots\mu_k}$, transforming as follows,

$$T'^{\mu_1\mu_2\cdots\mu_k} = \Lambda^{\mu_1}_{\ \nu_1}\Lambda^{\mu_2}_{\ \nu_2}\cdots\Lambda^{\mu_k}_{\ \nu_k}\, T'^{\nu_1\nu_2\cdots\nu_k}. \qquad (1.30)$$

Shifting indices is done in analogy to the vector components by means of the metric tensor,

$$T_{\mu}^{\ \nu} = g_{\mu\rho}T^{\rho\nu}, \quad T_{\mu\nu} = g_{\mu\rho}\, g_{\nu\sigma}\, T^{\rho\sigma}. \qquad (1.31)$$

In particular for the metric tensor the relations hold

$$g_{\mu\nu} = g^{\mu\nu}, \quad g_{\mu}^{\ \nu} = \delta^{\nu}_{\mu}, \quad g^{\mu}_{\ \nu} = \delta^{\mu}_{\nu} \qquad (1.32)$$

where $g^{\mu\nu}$ is defined via $g^{\mu\nu}x_{\mu}y_{\nu} = g_{\mu\nu}\, x^{\mu}y^{\nu}$ for all x and y.

1.3 Mechanics

The motion of a particle with mass m along a given trajectory is described by a time-dependent 3-dimensional vector $\vec{x}(t)$, with components $x^k(t)$, $k = 1, 2, 3$. The velocity is as usual defined as the derivative

$$\vec{v} = \frac{d\vec{x}}{dt} = (v^1, v^2, v^3), \quad v^k = \frac{dx^k}{dt}. \qquad (1.33)$$

The 4-dimensional trajectory is described by a curve in Minkowski space,

$$t \to x^{\mu}(t) \quad \text{with } x^0(t) = ct,$$

denoted as the *world line* of the particle. For neighbouring points x^{μ} und $x^{\mu} + dx^{\mu}$ along the world line the difference is given by

$$dx^{\mu} = \frac{dx^{\mu}}{dt}dt,$$

yielding the invariant 4-dimensional square, the line element of the world line,

$$ds^2 = g_{\mu\nu}\, dx^{\mu}dx^{\nu} = g_{\mu\nu}\,\frac{dx^{\mu}}{dt}\frac{dx^{\nu}}{dt}\, dt^2$$

$$= \left[\left(\frac{dx^0}{dt}\right)^2 - \vec{v}^2\right]dt^2 = c^2 dt^2\left[1 - \frac{\vec{v}^2}{c^2}\right] \equiv c^2\, d\tau^2. \qquad (1.34)$$

By this relation the invariant interval of the *proper time* is defined,

$$d\tau = dt \sqrt{1 - \frac{\vec{v}^2}{c^2}}. \qquad (1.35)$$

It is displayed by a co-moving clock.

4-velocity. The 4-velocity of a moving particle is defined by

$$u^\mu = \frac{dx^\mu}{d\tau} = \frac{1}{\sqrt{1 - \frac{\vec{v}^2}{c^2}}} \frac{dx^\mu}{dt}. \qquad (1.36)$$

By construction, this quantity is a 4-vector, with the covariant components

$$(u^\mu) = (u^0, \vec{u}) = \left(\frac{c}{\sqrt{1 - \frac{\vec{v}^2}{c^2}}}, \frac{\vec{v}}{\sqrt{1 - \frac{\vec{v}^2}{c^2}}} \right) \qquad (1.37)$$

and the 4-dimensional square

$$u^2 = (u^0)^2 - \vec{u}^2 = c^2.$$

4-momentum. The 4-momentum of a moving particle with mass m is defined by

$$p^\mu = m\, u^\mu. \qquad (1.38)$$

This is another 4-vector, with the square

$$p^2 = (p^0)^2 - \vec{p}^2 = m^2 c^2. \qquad (1.39)$$

For low velocities $|\vec{v}| \ll c$ one obtains the non-relativistic approximation

$$p^0 = \sqrt{\vec{p}^2 + m^2 c^2} = \sqrt{m^2 c^2 + \frac{m^2 \vec{v}^2}{1 - \frac{\vec{v}^2}{c^2}}} \approx mc + \frac{m\vec{v}^2}{2c} + \cdots \qquad (1.40)$$

and thus

$$c\, p^0 = mc^2 + \frac{m}{2}\vec{v}^2 + \cdots \equiv E, \qquad (1.41)$$

representing the energy of the particle. The first term is the rest energy given by the mass of the particle, the second term is the non-relativistic kinetic energy, and the higher terms in the expansion are relativistic corrections. The 4-momentum hence can be written as

$$(p^\mu) = (E/c, \vec{p}), \qquad (1.42)$$

and one finds the relativistic version of the energy–momentum correlation,

$$\boxed{E^2 = \vec{p}^2 c^2 + m^2 c^4}. \qquad (1.43)$$

Note that the mass m is a Lorentz invariant (scalar) quantity characterizing a particle.

1.3.1   *Covariant equation of motion*

The correct relativistic version of Newton's equation of motion can be formulated in terms of the 4-momentum and the proper time as follows,

$$\frac{dp^\mu}{d\tau} = K^\mu. \tag{1.44}$$

The quantity K^μ is a 4-vector, denoted as 4-force. Since all terms transform with the same matrix Λ, a Lorentz transformation into another reference frame, with $p^\mu \to p'^\mu$, $K^\mu \to K'^\mu$, yields the transformed equation of motion

$$\frac{dp'^\mu}{d\tau} = K'^\mu. \tag{1.45}$$

The individual components have changed, but the equation has the same form in different inertial frames; the equation is invariant in form, or *covariant*. In this context, *covariant* means that an equation can be written so that both sides have the same, well-defined, transformation properties under Lorentz transformations. This is an example for describing a physical law with the help of vector or tensor components, making Lorentz symmetry manifest.

The 3-dimensional part of Eq. (1.44) can be cast into the form

$$\frac{dp^k}{dt} = K^k\sqrt{1 - \frac{\vec{v}^2}{c^2}} \equiv F^k, \tag{1.46}$$

involving the 3-dimensional force $\vec{F}$; an example will be given below. The physical meaning of the 0-component of the 4-force follows from the indentity

$$p_\mu \frac{dp^\mu}{d\tau} = \frac{1}{2}\frac{dp^2}{d\tau} = 0 \quad (\text{since } p^2 = m^2c^2)$$

together with

$$p_\mu \frac{dp^\mu}{d\tau} = p_\mu K^\mu = p_0 K^0 - \vec{p}\vec{K} = 0$$

to become

$$p_0 K^0 = \vec{p}\vec{K}. \tag{1.47}$$

For $\vec{K} = 0$ ($\vec{F} = 0$) this implies $p^0 K^0 = 0$ and thus $K^0 = 0$ because of $p^0 > 0$. The 0-component of Eq. (1.44) in this case,

$$\frac{dp^0}{dt} = 0,$$

expresses the conservation of energy. For general $\vec{F}$ one obtains from Eq. (1.47),

$$\frac{d(cp^0)}{dt} = \vec{v} \cdot \vec{F}$$

describing the change of energy as the rate of work done by the force $\vec{F}$.

A concrete example for the 4-force is the electromagnetic force on a charged particle with charge e,

$$K^\mu = e\, F^{\mu\nu}\, u_\nu, \tag{1.48}$$

where $F^{\mu\nu}$ is the electromagnetic field-strength tensor (defined in Sec. 1.6). The 3-dimensional part of K^μ has the form

$$\vec{K} = \frac{e\vec{E} + e\vec{v} \times \vec{B}}{\sqrt{1 - \frac{\vec{v}^2}{c^2}}} \tag{1.49}$$

and the 3-dimensional part of the equation of motion (1.44) becomes

$$\frac{d\vec{p}}{dt} = \vec{K}\sqrt{1 - \frac{\vec{v}^2}{c^2}} = \vec{F} = e\,(\vec{E} + \vec{v} \times \vec{B}) \tag{1.50}$$

with the familiar expression $\vec{F}$ for the Lorentz force.

Convention

From now on we will use the system of *natural units*: velocities in units of c, action and angular momentum in units of $\hbar$, unit of energy is GeV (or MeV). Formally this means $\hbar = c = 1$. Conversion: $\hbar c = 197.329$ MeV fm. With natural units one has a simplified notation, for example

$$(x^\mu) = (t, \vec{x}), \quad (p^\mu) = (E, \vec{p}), \quad E^2 = \vec{p}^2 + m^2, \ldots$$

1.4 Lagrangian Formulation

Preliminary remark. In this section the usual notation for 3-dimensional vectors is exceptionally used, like $\vec{v} = (v_1, v_2, v_3)$, $\vec{p} = (p_1, p_2, p_3), \ldots$

1.4.1 *Free particle*

Consider the free motion of a particle with mass m, described by the coordinates x_i and the velocities $\dot{x}_i = v_i$. The corresponding Lagrangian is given by

$$L = -m\sqrt{1 - \vec{v}^2}. \tag{1.51}$$

The equations of motions follow as the Euler-Lagrange equations,

$$\frac{d}{dt}\left(\frac{\partial L}{\partial v_i}\right) - \frac{\partial L}{\partial x_i} = 0 \tag{1.52}$$

yielding

$$\frac{d}{dt}\left(\frac{mv_i}{\sqrt{1-\vec{v}^2}}\right) = 0, \qquad \frac{d\vec{p}}{dt} = 0. \tag{1.53}$$

They coincide with Eq. (1.46) for $\vec{F} = 0$ and confirm Eq. (1.51) as the correct relativistic version for the Lagrangian of a free particle.

Hamiltonian. The Hamiltonian H is derived from the Lagrangian L by means of a Legrendre transformation,

$$H = \sum_i p_i v_i - L \tag{1.54}$$

with the canonical momenta

$$p_i = \frac{\partial L}{\partial v_i}, \tag{1.55}$$

where the velocities v_i have to be substituted by the momenta p_i. Solving Eq. (1.55) for v_i one obtains

$$v_i = \frac{p_i}{\sqrt{\vec{p}^2 + m^2}},$$

and replacing all v_i in Eq. (1.54) yields the Hamiltonian

$$H = \sqrt{\vec{p}^2 + m^2} \tag{1.56}$$

corresponding to the energy of a free particle.

1.4.2 *Particle in an electromagnetic field*

The equations of motion (1.50) for a particle with charge e in an electromagnetic field can also be derived from a suitable Lagrangian. The Lagrangian method requires to describe the electromagnetic field in terms of the scalar potential ϕ and the vector potential $\vec{A}$, related to the field strengths according to

$$\vec{B} = \nabla \times \vec{A}, \quad \vec{E} = -\frac{\partial \vec{A}}{\partial t} - \nabla \phi. \tag{1.57}$$

The Lagrangian

$$L = -m\sqrt{1 - \vec{v}^2} + e\,\vec{v}\cdot\vec{A} - e\,\phi \tag{1.58}$$

yields the equations of motion as given in Eq. (1.50), as one can see from the following steps. The Euler-Lagrange equations (1.52)

$$\frac{d}{dt}\left(\frac{mv_i}{\sqrt{1-\vec{v}^2}}\right) + e\frac{dA_i}{dt} - \left(e\,\vec{v}\cdot\frac{\partial \vec{A}}{\partial x_i} - e\frac{\partial \phi}{\partial x_i}\right) = 0 \tag{1.59}$$

together with

$$\frac{dA_i}{dt} = \frac{\partial A_i}{\partial t} + \sum_k \frac{\partial A_i}{\partial x_k}\frac{dx_k}{dt} = \frac{\partial A_i}{\partial t} + \sum_k v_k \frac{\partial A_i}{\partial x_k} \tag{1.60}$$

result in

$$\frac{d}{dt}\left(\frac{mv_i}{\sqrt{1-\vec{v}^2}}\right) = e\left(-\frac{\partial A_i}{\partial t} - \frac{\partial \phi}{\partial x_i}\right) + e\sum_k v_k\left(\frac{\partial A_k}{\partial x_i} - \frac{\partial A_i}{\partial x_k}\right)$$

$$= e\left(\vec{E} + (\vec{v} \times \vec{B})\right)_i. \tag{1.61}$$

To identify the second term on the right of Eq. (1.61) as the magnetic force it is convenient to rewrite the force with the help of the ϵ-tensor,

$$(\vec{v} \times \vec{B})_i = \sum_{k,l} \epsilon_{ikl}\, v_k \sum_{m,n} \epsilon_{lmn} \frac{\partial A_n}{\partial x_m}$$

$$= \sum_{k,m,n}\left(\sum_l \epsilon_{ikl}\,\epsilon_{lmn}\right) v_k \frac{\partial A_n}{\partial x_m}$$

$$= \sum_{k,m,n}(\delta_{im}\delta_{kn} - \delta_{in}\delta_{km})\, v_k \frac{\partial A_n}{\partial x_m}$$

$$= \sum_k v_k\left(\frac{\partial A_k}{\partial x_i} - \frac{\partial A_i}{\partial x_k}\right),$$

which is exactly the expression appearing in Eq. (1.61).

Hamiltonian. The relativistic Hamiltonian that goes with the Lagrangian (1.58) is obtained by a Legendre transformation according to Eq. (1.54). The equations for the canonical momenta,

$$p_i = \frac{\partial L}{\partial v_i} = \frac{m v_i}{\sqrt{1 - \vec{v}^2}} + e\, A_i, \tag{1.62}$$

can be solved for v_i yielding the velocities in terms of the momenta,

$$v_i = \frac{p_i - e A_i}{\sqrt{\left(\vec{p} - e\vec{A}\right)^2 + m^2}}.$$

Substituting all v_i by p_i in Eq. (1.54) leads to the Hamiltonian

$$H = e\,\phi + \sqrt{\left(\vec{p} - e\vec{A}\right)^2 + m^2}. \tag{1.63}$$

This expression is recognized to follow from the Hamiltonian (1.56) for a free particle by the substitution

$$\boxed{\vec{p} \to \vec{p} - e\,\vec{A}, \quad H \to H - e\,\phi} \tag{1.64}$$

denoted as *minimal substitution*, which plays a fundamental role in the description of the electromagnetic interaction at the classical and at the quantum level.

1.5   Fields and Derivatives

So far we have encountered scalars, vectors, and tensors as quantities with different well defined properties under Lorentz transformations. Further concepts required in relativistic field theories are fields and their partial derivatives. In general, a field φ is defined as a function of the four space-time variables x^μ,

$$\varphi(x^0, x^1, x^2, x^3) \equiv \varphi(x).$$

According to the properties under Lorentz transformations several classes of fields have to be distinguished.

1.5.1   *Scalar field*

A scalar field $\phi(x)$ is invariant under a Lorentz transformation Λ,

$$\Lambda: \quad \phi(x) \to \phi'(x') = \phi(x) = \phi(\Lambda^{-1} x') \tag{1.65}$$

with $x^\mu \to x'^\mu$ given in Eq. (1.10), in short denoted by $x' = \Lambda x$.

1.5.2 *Vector field*

A vector field $A^\mu(x)$ transforms as a 4-vector,

$$\Lambda: \quad A^\mu(x) \to A'^\mu(x') = \Lambda^\mu{}_\nu A^\nu(x), \tag{1.66}$$

involving transformations of both components and variables.

1.5.3 *Tensor field*

A tensor field $T^{\mu\nu}$ is characterized by the property that each index transforms like a vector index, e.g. for a second-rank tensor field one has

$$\Lambda: \quad T^{\mu\nu}(x) \to T'^{\mu\nu}(x') = \Lambda^\mu{}_\rho \Lambda^\nu{}_\sigma T^{\rho\sigma}(x). \tag{1.67}$$

Defining tensor fields $T^{\mu_1\mu_2\cdots\mu_k}(x)$ of higher rank is straight forward. Vectors and scalars can be considered as tensors of rank one and zero.

1.5.4 *Partial derivatives*

For the partial derivatives with respect to x^μ and x'^μ a short notation is introduced,

$$\partial_\mu = \frac{\partial}{\partial x^\mu}, \quad \partial'_\mu = \frac{\partial}{\partial x'^\mu}. \tag{1.68}$$

These derivatives transform under Lorentz transformations as covariant 4-vector components according to Eq. (1.26),

$$\partial_\nu = \frac{\partial}{\partial x^\nu} = \left(\frac{\partial}{\partial x'^\mu}\right)\frac{\partial x'^\mu}{\partial x^\nu} = \partial'_\mu \Lambda^\mu{}_\nu. \tag{1.69}$$

The 4-vector with the covariant components ∂_μ is the 4-dimensional gradient, also denoted as *4-gradient*. The contravariant components follow as usual by means of the metric tensor,

$$\partial^\mu = g^{\mu\nu}\partial_\nu, \quad (\partial^\mu) = \left(\frac{\partial}{\partial t}, -\nabla\right). \tag{1.70}$$

A Lorentz-invariant differential operator is given by the square of the 4-gradient, representing the d'Alembert operator,

$$\partial_\mu \partial^\mu = \frac{\partial^2}{\partial t^2} - \Delta = \Box. \tag{1.71}$$

Differentiation raises the rank of tensor fields,

$$\begin{array}{lll} \text{scalar field } \phi & \Rightarrow & \partial_\mu \phi \quad \text{vector field} \\ \text{vector field } A_\nu & \Rightarrow & \partial_\mu A_\nu \quad \text{tensor field} \\ \text{tensor field } T_{\nu\rho} & \Rightarrow & \partial_\mu T_{\nu\rho} \quad \text{tensor field of rank 3} \\ etc. \end{array}$$

1.6 Electrodynamics

This section contains the covariant formulation of electrodynamics which is invariant in form under Lorentz transformations.

Since electrodynamics is already a Lorentz-invariant theory, it is not necessary to modify the formulation of the physical laws. On the other hand, Lorentz symmetry is hidden when the conventional 3-dimensional notation is taken. Instead, the use of 4-dimensional vectors and tensors facilitates a covariant formulation of Maxwell's equations making the symmetry under Lorentz transformations manifest.

1.6.1 *Potentials*

The electromagnetic potentials ϕ und $\vec{A}$, related to the field strengths $\vec{B}$ and $\vec{E}$ by means of Eq. (1.57), can be summarized in terms of a vector field, the 4-potential

$$(A^\mu) = (A^0, \vec{A}) = (\phi, \vec{A}), \quad (A_\mu) = (\phi, -\vec{A}). \tag{1.72}$$

1.6.2 *Field strengths*

The field-strength tensor follows as the antisymmetric derivative of the 4-potential,

$$F_{\mu\nu} = \partial_\mu A_\nu - \partial_\nu A_\mu, \quad F^{\mu\nu} = \partial^\mu A^\nu - \partial^\nu A^\mu. \tag{1.73}$$

Being antisymmetric, $F_{\mu\nu} = -F_{\nu\mu}$, this tensor has 6 independent components, related to the field strengths as follows,

$$\vec{E} = (E_x, E_y, E_z) = (F_{01}, F_{02}, F_{03}),$$
$$\vec{B} = (B_x, B_y, B_z) = (-F_{23}, -F_{31}, -F_{12}), \tag{1.74}$$

as one can easily derive from Eq. (1.57). Thus one can write

$$(F_{\mu\nu}) = \begin{pmatrix} 0 & E_x & E_y & E_z \\ -E_x & 0 & -B_z & B_y \\ -E_y & B_z & 0 & -B_x \\ -E_z & -B_y & B_x & 0 \end{pmatrix}. \tag{1.75}$$

Since the entries are not the spatial components of 4-vectors, they are labeled as 3-dimensional x, y, z-components.

1.6.3 *Gauge transformations*

The potentials are not unique, they are determined only up to gauge transformations

$$A_\mu \to A_\mu + \partial_\mu \chi \tag{1.76}$$

with arbitrary functions $\chi(x)$, leaving the field strengths (1.73) invariant
because the partial derivatives commute. A restriction of this gauge degree
of freedom is obtained by specifying a gauge condition. Of particular interest is the *Lorentz gauge*

$$\partial_\mu A^\mu = 0, \tag{1.77}$$

which constitutes a relativistic-invariant condition, valid in all inertial
frames.

1.6.4 *Electromagnetic current*

The charge and current densities ρ and $\vec{j}$ can be summarized in terms of a
4-vector, the *4-current*,

$$(j^\mu) = (j^0, \vec{j}) = (\rho, \vec{j}). \tag{1.78}$$

The current fulfills a continuity equation

$$\frac{\partial \rho}{\partial t} + \nabla \cdot \vec{j} = 0, \tag{1.79}$$

which corresponds to current conservation. In 4-dimensional notation the
continuity equation is a vanishing 4-divergence,

$$\partial_\mu j^\mu = 0, \tag{1.80}$$

and obviously invariant under Lorentz transformations. The total charge

$$Q = \int d^3x\, \rho = \int d^3x\, j^0 \tag{1.81}$$

is a Lorentz invariant and, as a consequence of the continuity equation, also
a conserved quantity.

1.6.5 *Maxwell's equations*

With the help of the field-strength tensor the set of Maxwell's equations
can be formulated in a covariant way as follows,

$$\partial_\mu F^{\mu\nu} = j^\nu \quad \text{inhomogeneous equations} \tag{1.82}$$

$$\epsilon^{\mu\nu\rho\sigma} \partial_\nu F_{\rho\sigma} = 0 \quad \text{homogeneous equations.} \tag{1.83}$$

The homogeneous equations are solved by expressing the field strengths in terms of the potentials. Insertion of Eq. (1.73) into the inhomogeneous equations yields coupled equations for the components of the 4-potential,

$$\partial_\mu \left(\partial^\mu A^\nu - \partial^\nu A^\mu\right) = \Box A^\nu - \partial^\nu \left(\partial_\mu A^\mu\right) = j^\nu. \tag{1.84}$$

Choosing the Lorentz gauge for A^μ simplifies these equations considerably and one obtains decoupled equations,

$$\Box A^\nu = j^\nu. \tag{1.85}$$

This set of inhomogeneous wave equations with the constraint (1.77) can be considered equivalent to the set of Maxwell's equations.

1.6.6 *Energy-momentum tensor*

The field strengths allow to construct a tensor of second rank that comprises the mechanical properties energy and momentum of the electromagnetic field,

$$T^{\mu\nu} = F^\mu_{\ \rho} F^{\rho\nu} + \frac{1}{4} g^{\mu\nu} \left(F_{\rho\sigma} F^{\rho\sigma}\right). \tag{1.86}$$

This energy-momentum tensor contains the energy density of the field,

$$T^{00} = \frac{1}{2} \left(\vec{E}^2 + \vec{B}^2\right) \tag{1.87}$$

and the components of the Poynting vector $\vec{S} = \vec{E} \times \vec{B}$,

$$S^k = T^{0k} \tag{1.88}$$

representing the momentum density. The purely spatial components T^{kl}, also known as Maxwell's stress tensor, describe the momentum-current density of the electromagnetic field.

The force on a charge and current distribution j^μ is given by a generalization of Eq. (1.50) to continuous systems,

$$F^k = \int d^3x \left(\rho \vec{E} + \vec{j} \times \vec{B}\right)^k = \int d^3x \, F^{k\nu} j_\nu \equiv \int d^3x \, f^k(x) \tag{1.89}$$

with the force density f^k that can be extended to a 4-vector f^μ and written in a covariant way as follows,

$$f^\mu = F^{\mu\nu} j_\nu. \tag{1.90}$$

Using Maxwell's equations (1.82) to replace j_ν, and (1.83) to rearrange the various derivative terms, one obtains the following relation to the energy-momentum tensor,

$$f^\mu = -\partial_\nu T^{\mu\nu}. \tag{1.91}$$

For $j_\nu = 0$, one has a free radiation field which fulfills

$$\partial_\nu T^{\mu\nu} = 0, \tag{1.92}$$

that is the conservation law for energy and momentum of the free field (see also Sec. 4.2 for more details). The general expression (1.91) represents energy and momentum conservation, including the rate of work done by the field on the charges of the current.

Chapter 2

Elements of Relativistic Quantum Field Theory

The quantum mechanical description of particles in accordance with the principles of relativity, relativistic quantum mechanics, turns out to be a problematic concept with fundamental difficulties that are finally based on the equivalence of mass and energy resulting in non-conserved particle numbers.

Starting the relativistic description of a particle with mass m in the context of quantum mechanics the usual procedure follows the correspondence principle, the replacement of classical observables by quantum mechanical operators acting on the space of states of the particle, according to the scheme

classical observable	$\rightarrow$	quantum mechanical operator
momentum $\vec{p}$	$\rightarrow$	momentum operator $\quad -i\nabla$ $P^k = -i\partial_k = i\partial^k$
energy p^0	$\rightarrow$	$i\frac{\partial}{\partial t} = i\partial_0 = i\partial^0$
4-momentum	$\rightarrow$	$P^\mu = i\partial^\mu, \quad P_\mu = i\partial_\mu$
Hamiltonian	$\rightarrow$	Hamiltonian operator

The square of 4-momentum corresponds to the invariant operator

$$P_\mu P^\mu = -\partial_\mu \partial^\mu = -\Box. \qquad (2.1)$$

Commutation relations for position and momentum operators are still valid,

$$\left[X^k, P^l\right] = i\,\delta^{kl}. \tag{2.2}$$

However, already for a free particle one encounters a problem. Replacing the classical Hamiltonian via correspondence

$$H = \sqrt{\vec{p}^2 + m^2} \to \sqrt{-\Delta + m^2}$$

does not provide a local differential operator, and the quantum mechanical equation of motion

$$i\,\frac{\partial\phi}{\partial t} = H\phi \tag{2.3}$$

for the wave function ϕ of a particle state is not a differential equation. Two different roads were proposed to prevent the problem, one of them leads to the Klein–Gordon equation and the other one to the Dirac equation.

2.1 Klein–Gordon Equation

A first attempt for treating spinless relativistic particles in the frame of one-particle quantum mechanics is based on the Klein–Gordon equation. The emerging problems can be solved by a transition from quantum mechanics to quantum field theory.

2.1.1 *Relativistic quantum mechanics of spin-0 particles*

The difficulty with the Hamiltonian as the square root of a differential operator can be circumvented by a twofold application of the operators on both sides of Eq. (2.3), yielding a differential equation of second order,

$$i\,\frac{\partial}{\partial t}\left(i\,\frac{\partial\phi}{\partial t}\right) = H^2\phi \quad\Leftrightarrow\quad -\frac{\partial^2}{\partial t^2}\phi = \left(-\Delta + m^2\right)\phi. \tag{2.4}$$

It can be used to replace Eq. (2.3) as the quantum mechanical equation of motion for the wave function $\phi = \phi(t, \vec{x})$ of a particle state,

$$\boxed{\left(\Box + m^2\right)\phi = 0}. \tag{2.5}$$

This differential equation, which is of second order also with respect to time, is known as the *Klein–Gordon equation* (although it was considered also by Schrödinger). Because of the invariant differential operator, it is a Lorentz-invariant equation when $\phi(x)$ is a scalar field under Lorentz transformations.

In the non-relativistic interpretation, $|\phi|^2 = \rho$ is the probability density of the particle's position in space with the normalization

$$\int d^3x \, |\phi|^2 = 1 \qquad (2.6)$$

in accordance with the number of particles as a conserved quantity. In a relativistic version this interpretation cannot be maintained since the volume element d^3x is not invariant and thus the normalization would be dependent on the reference frame.

An analogy to the non-relativistic quantities, probability density ρ and probability current $\vec{j}$, is found in terms of the 4-current associated with the Klein–Gordon wave function ϕ,

$$j^\mu = i \left[\phi^* \partial^\mu \phi - (\partial^\mu \phi^*) \phi \right]. \qquad (2.7)$$

This current is conserved,

$$\partial_\mu j^\mu = 0, \qquad (2.8)$$

as can be easily derived from the Klein–Gordon equation and its Hermitian adjoint,

$$\phi^* \left(\Box + m^2 \right) \phi - \phi \left(\Box + m^2 \right) \phi^* = 0$$
$$= \partial_\mu \left(\phi^* \partial^\mu \phi - \phi \, \partial^\mu \phi^* \right).$$

The continuity equation (2.8) implies the existence of a Lorentz-invariant conserved quantity,

$$\frac{d}{dt} \int d^3x \, j^0 = 0. \qquad (2.9)$$

Nevertheless, the interpretation of $\rho = j^0$ as a probability density is not possible because j^0 is not positive definite . This can be seen immediately for the example of plane waves as solutions of Eq. (2.5),

$$\phi_\pm = e^{\mp ipx} = e^{\mp i(Et - \vec{p}\vec{x})} \quad \text{with } E = p^0 = \sqrt{\vec{p}^2 + m^2}, \qquad (2.10)$$

yielding $j^0_\pm = \pm 2E$.

In the correct interpretation, the Klein–Gordon equation decribes two kinds of particles with opposite charges ± 1 (in a suitable normalization): particles $(+1)$ and antiparticles (-1), both with the same mass m. The plane waves (2.10) can be considered to be the wave functions of particles (ϕ_+) and of antiparticles (ϕ_-) with momentum $\vec{p}$ and energy E. The appropriate framework is provided by relativistic quantum field theory.

2.1.2 *Quantum field theoretical formulation*

The states of both particles and antiparticles are combined in a common Hilbert space. The eigenstates for momentum and energy of a particle or antiparticle with the eigenvalues

$$\vec{p} \quad \text{and} \quad p^0 = \sqrt{\vec{p}^2 + m^2}$$

are described by ket vectors in the conventional quantum mechanical language,

$$\text{particle state} \quad |+, p\rangle$$
$$\text{antiparticle state} \quad |-, p\rangle \tag{2.11}$$

Furthermore, there is a zero-particle state,

$$\text{vacuum} \quad |0\rangle \tag{2.12}$$

with the normalization

$$\langle 0|0\rangle = 1. \tag{2.13}$$

The momentum eigenstates as part of a continuous spectrum can only be normalized in terms of a δ-function,

$$\langle \pm, p \,|\, \pm, p'\rangle = 2p^0 \, \delta^3(\vec{p} - \vec{p}') \tag{2.14}$$
$$\langle \pm, p \,|\, \mp, p'\rangle = 0.$$

This differs from the standard normalization in quantum mechanics by a factor $2p^0$. The choice of Eq. (2.14) has the advantage of being invariant under Lorentz transformations,

$$\int \frac{d^3p}{2p^0} \, \langle \pm, p \,|\, \pm, p'\rangle = 1 \tag{2.15}$$

with the invariant volume element $d^3p/(2p^0)$. The last-mentioned property follows from the invariance of the 4-dimensional volume element d^4p according to

$$\int d^4p \, \delta(p^2 - m^2) \, \theta(p^0) = \int d^3p \underbrace{\int dp^0 \, \delta(p^{0\,2} - \vec{p}^2 - m^2) \, \theta(p^0)}.$$

$$= \frac{1}{2p^0} \quad \text{with } p^0 = \sqrt{\vec{p}^2 + m^2}$$

More-particle states are obtained by the usual procedure as products of one-particle states and subsequent symmetrization. In this way the space of states is constructed as a multi-particle state, without a definite number of particles. This reflects the physical situation that particle numbers cannot be conserved quantitites since particles can be annnihilated, can be created, can be converted ... as far as fundamental conservation laws are not violated.

The space of states as an algebraic quantity is accompanied by a field quantity, the solution of the Klein–Gordon equation, which in the framework of a quantum field theory becomes a *field operator*, also called a *quantum field*, acting as a linear operator on the space of states of particles and antiparticles. The general solution of the Klein–Gordon equation (2.5) can be written as a Fourier expansion in terms of plane waves,

$$\boxed{\phi(x) = \frac{1}{(2\pi)^{3/2}} \int \frac{d^3 p}{2p^0} \left[a(p) \, e^{-ipx} + b^\dagger(p) \, e^{ipx} \right]} \tag{2.16}$$

where once again the invariant volume element is used. In case of a classical solution the coefficents in the expansion (2.16) would be complex numbers. With $\phi(x)$ as a quantum field they are operators acting on the space of states, namely as

$$a, b : \quad \text{annihilation operators}$$

$$a^\dagger, b^\dagger : \quad \text{creation operators}$$

of particles (a) and antiparticles (b). Their detailed mode of operation is defined as follows,

$$a^\dagger(p) \, |0\rangle = |+, p\rangle, \quad b^\dagger(p) \, |0\rangle = |-, p\rangle$$

$$a(p) \, |+, p'\rangle = 2p^0 \, \delta^3 \left(\vec{p} - \vec{p}' \right) |0\rangle$$

$$b(p) \, |-, p'\rangle = 2p^0 \, \delta^3 \left(\vec{p} - \vec{p}' \right) |0\rangle \tag{2.17}$$

$$a(p) \, |-, p'\rangle = b(p) \, |+, p'\rangle = 0$$

The annihilation and creation operators fulfill canonical commutation relations, in analogy to the lowering and raising operators for the harmonic oscillator,

$$\left[a(p), a(p') \right] = \left[b(p), b(p') \right] = 0$$

$$\left[a(p), a^\dagger(p') \right] = 2p^0 \, \delta^3 \left(\vec{p} - \vec{p}' \right)$$

$$\left[b(p), b^\dagger(p') \right] = 2p^0 \, \delta^3 \left(\vec{p} - \vec{p}' \right) \tag{2.18}$$

$$\left[a(p), b(p') \right] = \left[a(p), b^\dagger(p') \right] = 0$$

(the residual combinations follow via Hermitian conjugation). Differences with respect to the harmonic oscillator are

- a pair of oscillators for each $\vec{p}$ with frequency $\omega = p^0$,
- continuous index $\vec{p}$,
- relativistic normalization.

Wave functions of one-particle states appear as matrix elements of the field operator between the vacuum and the one-particle states,

$$\langle 0 \,|\, \phi(x) \,|\, +, p\rangle = \langle 0 \,|\, \phi^\dagger(x) \,|\, -, p\rangle = \frac{1}{(2\pi)^{3/2}}\, e^{-ipx},$$

$$\langle -, p \,|\, \phi(x) \,|\, 0\rangle = \langle +, p \,|\, \phi^\dagger(x) \,|\, 0\rangle = \frac{1}{(2\pi)^{3/2}}\, e^{ipx}.$$

(2.19)

They enter the description of processes where particles/ antiparticles are created or annihilated.

2.1.3 *Current and charge*

With the field operator ϕ the current (2.7) becomes a current operator

$$j^\mu = i \left[\phi^\dagger \partial^\mu \phi - (\partial^\mu \phi^\dagger)\phi \right]$$

(2.20)

with a conserved charge following from the continuity equation (2.8),

$$Q = \int d^3x\, j^0.$$

(2.21)

By insertion of the Fourier expansion for ϕ in j^0 one can express the charge operator in terms of the number operators $N_+(p)$ and $N_-(p)$ for particles and antiparticles as follows,

$$Q = \int \frac{d^3p}{2p^0}\, [a^\dagger(p)a(p) - b^\dagger(p)b(p)] \equiv \int \frac{d^3p}{2p^0}\, [N_+(p) - N_-(p)].$$

(2.22)

One can easily verify by means of Eq. (2.17) that the one-particle states are eigenstates of Q with eigenvalues ± 1:

$$Q\,|+, p\rangle = +\,|+, p\rangle, \quad Q\,|-, p\rangle = -\,|-, p\rangle.$$

(2.23)

A special case is the real (self-adjoint) field with $\phi(x) = \phi^\dagger(x)$, which describes neutral particles of charge zero where particles and antiparticles are identical.

2.1.4 *Mechanical observables*

The mechanical observables for the system described by the quantum field (2.16) can be expressed in terms of the number operators, in a similar way as done for the charge,

$$\text{energy}\quad H = \int \frac{d^3p}{2p^0}\, p^0 \left[N_+(p) + N_-(p) \right],$$

$$\text{momentum}\quad \vec{P} = \int \frac{d^3p}{2p^0}\, \vec{p} \left[N_+(p) + N_-(p) \right].$$

(2.24)

With the help of the rules (2.17) one can easily verify that the one-particle states are eigenstates,

$$H \left| \pm, p \right\rangle = p^0 \left| \pm, p \right\rangle, \quad \vec{P} \left| \pm, p \right\rangle = \vec{p} \left| \pm, p \right\rangle. \tag{2.25}$$

For a neutral field the expressions (2.24) are also valid, with only a single number operator $N(p)$ instead of $N_+(p) + N_-(p)$.

The representation of energy and momentum of the field in terms of the number operators and the eigenvalues of the one-particle states is plausible from a physics point of view. A deeper reason for the expressions (2.24) and a representation in terms of the field operator $\phi(x)$ can be given within the Lagrangian formulation of field theory. For details we refer to Sec. 4.2.1.

Remark on normal ordering. The evaluation of integrals over bilinear expressions of field operators, as for the various observables, leads to unwanted constants which are formally divergent. Being properties of the vauum, they are unphysical and have to be subtracted to get the vacuum with quantum numbers zero. This subtraction can be organized by a procedure called normal ordering, indicated by placing the product of field operators between colons. The symbol $:\ldots:$ is an instruction to order all products such that creation operators appear to the left of annihilation operators; in case of fermion operators an extra minus sign has to be allocated for each commutation needed for normal ordering. We will not make use of the notation with colons and always assume that normal ordering is applied where needed.

2.2 Dirac Equation

The Klein–Gordon equation for a scalar field is suited for the description of particles without spin. On the other hand, fundamental particles with spin $1/2$, like the electron and the heavier leptons μ and τ (as well as quarks) are described by the Dirac equation.

2.2.1 *Relativistic quantum mechanics of spin-1/2 particles*

As a first step we consider the formulation of relativistic quantum mechanics for spin-1/2 particles. Starting point is the equation of motion (2.3) with the first order time derivative, but with an ansatz for H linear in the momentum,

$$H = \underbrace{\sum_{k=1}^{3} \alpha^k P^k}_{\vec{\alpha} \cdot \vec{P}, \quad \vec{\alpha} = (\alpha^1, \alpha^2, \alpha^3)} + \beta \, m$$

(2.26)

which has to fulfill the condition

$$H^2 = \vec{P}^2 + m^2 = \sum_k (P^k)^2 + m^2.$$

Comparison with

$$H^2 = \sum_{k,l} \frac{1}{2} (\alpha^k \alpha^l + \alpha^l \alpha^k) P^k P^l + \sum_k (\alpha^k \beta + \beta \alpha^k) P^k \, m + \beta^2 \, m^2$$

yields a set of equations for the quantitites α^k und β:

$$\alpha^k \alpha^l + \alpha^l \alpha^k = 2 \, \delta^{kl},$$
$$\alpha^k \beta + \beta \alpha^k = 0, \quad \beta^2 = 1.$$

(2.27)

This conditions cannot be fulfilled by numbers but require matrices that in addition have to be Hermitian in order to make H in (2.26) a self-adjoint operator. The rules (2.27) imply

- $\beta^2 = 1, \quad (\alpha^k)^2 = 1$
- eigenvalues of α^k and β are ± 1
- $\mathrm{Tr}(\beta) = \mathrm{Tr}(\alpha^k) = 0$.

These properties require an even number of dimensions D, with the lowest possible dimension given by $D = 4$. The quantities α^k, β hence are 4×4-matrices. They can be specified in a concrete way, written in terms of block submatrices as follows,

$$\beta = \begin{pmatrix} \mathbb{1} & 0 \\ 0 & -\mathbb{1} \end{pmatrix}, \quad \alpha^k = \begin{pmatrix} 0 & \sigma^k \\ \sigma^k & 0 \end{pmatrix}.$$

(2.28)

The entries are in each case 2×2 blocks with the respective zero and unit matrices; σ^k denote the *Pauli matrices*

$$\sigma^1 = \begin{pmatrix} 0 & 1 \\ 1 & 0 \end{pmatrix}, \quad \sigma^2 = \begin{pmatrix} 0 & -i \\ i & 0 \end{pmatrix}, \quad \sigma^3 = \begin{pmatrix} 1 & 0 \\ 0 & -1 \end{pmatrix}. \tag{2.29}$$

The representation (2.28) is known as the *Dirac representation* or *standard representation*. It is not the only possibility because the matrices are determined by the relations (2.27) only up to an equivalence transformation,

$$\alpha^k \rightarrow T\alpha^k T^{-1}, \quad \beta \rightarrow T\beta T^{-1}. \tag{2.30}$$

With Eq. (2.28) the equation of motion for a wave function ψ is given by

$$\boxed{i\frac{\partial \psi}{\partial t} = (\vec{\alpha}\cdot\vec{P} + \beta m)\psi} \tag{2.31}$$

which is the *Dirac equation*. Since it is a matrix equation, the wave function is a 4-component column,

$$\psi = \begin{pmatrix} \psi_1 \\ \psi_2 \\ \psi_3 \\ \psi_4 \end{pmatrix} \tag{2.32}$$

denoted as a *Dirac spinor*; each component is a function of the space-time variables, thus $\psi_a = \psi_a(x)$ for $a = 1, \ldots 4$.

2.2.1.1 *Observables*

The observables momentum and energy are represented by the operators

$$\vec{P} = -i\,\nabla,$$
$$H = \vec{\alpha}\cdot\vec{P} + \beta\,m = -i\,\vec{\alpha}\cdot\nabla + \beta\,m. \tag{2.33}$$

Angular momentum is composed of two contributions, orbital angular momentum $\vec{L}$ and spin $\vec{S}$. The operator for orbital angular momentum reads as usual

$$\vec{L} = \vec{X} \times \vec{P} = (L^1, L^2, L^3), \tag{2.34}$$

and the spin operator is given by the matrices

$$\vec{S} = \frac{1}{2}\vec{\Sigma} = (S^1, S^2, S^3) \tag{2.35}$$

with

$$\vec{\Sigma} = \left(\Sigma^1, \Sigma^2, \Sigma^3\right), \quad \Sigma^k = \begin{pmatrix} \sigma^k & 0 \\ 0 & \sigma^k \end{pmatrix}. \tag{2.36}$$

The following properties ensure that $\vec{S}$ indeed represents spin $1/2$:

$$\left[S^k, S^l\right] = i\epsilon_{klm}\, S^m,$$

$$\vec{S}^2 = \frac{3}{4}\mathbf{1} = s(s+1)\mathbf{1} \quad \text{with } s = \frac{1}{2},$$

$$S^3 = \frac{1}{2}\begin{pmatrix} 1 & 0 & 0 & 0 \\ 0 & 1 & 0 & 0 \\ 0 & 0 & -1 & 0 \\ 0 & 0 & 0 & -1 \end{pmatrix}$$

with the eigenvalues $m_s = \pm 1/2$.

From Eqs. (2.33)–(2.36) one can easily derive the commutation relations

$$\left[H, \vec{L}\right] = -\left[H, \vec{S}\right] = -i\,\vec{\alpha} \times \vec{P} \neq 0$$

which imply that orbital angular momentum and spin are not separately conserved and the eigenvalues m_s of S^3 are no good quantum numbers in general. On the other hand, the total angular momentum $\vec{J} = \vec{L} + \vec{S}$ is a conserved quantity,

$$\left[H, \vec{J}\right] = \left[H, \vec{L}\right] + \left[H, \vec{S}\right] = 0. \tag{2.37}$$

This shows that the expression for H in Eq. (2.33) contains an intrinsic spin–orbit interaction. Because of

$$\vec{P} \cdot \vec{J} = \vec{P} \cdot \left(\vec{X} \times \vec{P}\right) + \vec{P} \cdot \vec{S} = \vec{P} \cdot \vec{S}$$

and

$$\left[\vec{P} \cdot \vec{J}, H\right] = \vec{P} \cdot \left[\vec{J}, H\right] + \left[\vec{P}, H\right] \cdot \vec{J} = 0$$

the product $\vec{P} \cdot \vec{S}$ is also conserved. For momentum eigenstates the momentum operator can be replaced by the eigenvalue $\vec{p}$,

$$\vec{P} \cdot \vec{S} = \vec{p} \cdot \vec{S}.$$

Hence, the projection of spin on the direction of momentum, the *helicity*,

$$\frac{\vec{p}}{|\vec{p}|} \cdot \vec{S} \equiv \vec{n} \cdot \vec{S} = \frac{1}{2}\,\vec{n} \cdot \vec{\Sigma} \tag{2.38}$$

is a conserved quantity with eigenvalues $\pm 1/2$, which are good quantum numbers. Particle states thus can be determined as common eigenstates of the commuting observables H, $\vec{P}$, $\vec{n} \cdot \vec{S}$ by specifying the eigenvalues of momentum and helicity.

2.2.1.2 *Dirac matrices*

The matrices α^k and β in the Dirac Hamiltonian can be expressed in a covariant way by the redefinition

$$\gamma^0 = \beta, \quad \gamma^k = \beta \alpha^k \tag{2.39}$$

and can be combined in a formal 4-vector according to

$$\left(\gamma^\mu\right) = \left(\gamma^0, \gamma^1, \gamma^2, \gamma^3\right) = \left(\gamma^0, \vec{\gamma}\right). \tag{2.40}$$

The relations (2.27) become more compact when written as anti-commutation relations in terms of the *Dirac matrices* γ^μ,

$$\boxed{\gamma^\mu \gamma^\nu + \gamma^\nu \gamma^\mu \equiv \{\gamma^\mu, \gamma^\nu\} = 2\, g^{\mu\nu}\, \mathbb{1}} \tag{2.41}$$

where the symbol $\{\,,\,\}$ indicates the anti-commutator. The relations (2.41) carry the name *Dirac algebra*. Immediate consequences are

$$\left(\gamma^0\right)^2 = \mathbb{1}, \quad \left(\gamma^k\right)^2 = -\mathbb{1}. \tag{2.42}$$

By means of the metric tensor Dirac matrices with lower indices can also be defined:

$$\gamma_\mu = g_{\mu\nu}\gamma^\nu, \quad \text{i.e. } \gamma^0 = \gamma_0, \quad \gamma^k = -\gamma_k. \tag{2.43}$$

The set γ^μ is augmented by a further Dirac matrix defined as follows,

$$\gamma_5 = i\,\gamma^0 \gamma^1 \gamma^2 \gamma^3 \tag{2.44}$$

with the properties

$$\gamma_5^2 = \mathbb{1}, \quad \{\gamma_5, \gamma^\mu\} = 0. \tag{2.45}$$

In the Dirac representation specified in Eq. (2.28) the Dirac matrices are given by

$$\gamma^0 = \begin{pmatrix} \mathbb{1} & 0 \\ 0 & -\mathbb{1} \end{pmatrix}, \quad \gamma^k = \begin{pmatrix} 0 & \sigma^k \\ -\sigma^k & 0 \end{pmatrix}, \quad \gamma_5 = \begin{pmatrix} 0 & \mathbb{1} \\ \mathbb{1} & 0 \end{pmatrix}. \tag{2.46}$$

Other representations are obtained by equivalence transformations

$$\gamma^\mu \to T\gamma^\mu T^{-1} \tag{2.47}$$

which leave the Dirac algebra (2.41) unchanged. Of special importance for symmetry considerations, in particular in the context of Lorentz symmetry, is the *chiral representation*, also called *Weyl representation*. In the chiral representation the Dirac matrices read as follows,

$$\gamma^0 = \begin{pmatrix} 0 & \mathbb{1} \\ \mathbb{1} & 0 \end{pmatrix}, \quad \gamma^k = \begin{pmatrix} 0 & \sigma^k \\ -\sigma^k & 0 \end{pmatrix}, \quad \gamma_5 = \begin{pmatrix} -\mathbb{1} & 0 \\ 0 & \mathbb{1} \end{pmatrix}. \tag{2.48}$$

They are obtained from the Dirac representation via the transformation (2.47) with the matrix

$$T = \frac{1}{\sqrt{2}} \begin{pmatrix} \mathbb{1} & -\mathbb{1} \\ \mathbb{1} & \mathbb{1} \end{pmatrix}. \tag{2.49}$$

2.2.1.3 *Covariant version of the Dirac equation*

Starting from the Dirac equation (2.31) with $\vec{\alpha}\cdot\vec{P} = -i\alpha^k \partial_k$ one obtains by multiplication with β,

$$i\beta\partial_0\psi = -i\beta\alpha^k\partial_k\psi + \beta^2 m\psi$$

and by means of Eq. (2.39)

$$i\gamma^0\partial_0\psi + i\gamma^k\partial_k\psi - m\psi = i\gamma^\mu\partial_\mu\psi - m\psi = 0,$$

yielding the covariant version of the Dirac equation

$$\boxed{\left(i\gamma^\mu\partial_\mu - m\right)\psi = 0}. \tag{2.50}$$

The prove of covariance, i.e. validity of this form in any inertial frame, requires a detailed discussion of the behaviour of Dirac spinors under Lorentz transformations and is postponed to Sec. 2.3.

The adjoint spinor is defined by

$$\overline{\psi} = \psi^\dagger\gamma^0, \quad \psi^\dagger = \left(\psi_1^*, \psi_2^*, \psi_3^*, \psi_4^*\right). \tag{2.51}$$

Hermitian conjugation of Eq. (2.50) yields the *adjoint Dirac equation*

$$i\left(\partial\overline{\psi}\right)\gamma^\mu + m\overline{\psi} = 0 \tag{2.52}$$

by the use of

$$\overline{\gamma}^\mu \equiv \gamma^0\gamma^{\mu\dagger}\gamma^0 = \gamma^\mu. \tag{2.53}$$

In general one defines for arbitrary Dirac matrices and their products:

$$\overline{\Gamma} \equiv \gamma^0\Gamma^\dagger\gamma^0. \tag{2.54}$$

2.2.1.4 *Conserved current*

From the Dirac equation (2.50) and the adjoint Dirac equation (2.52) one can derive a conserved current

$$\boxed{j^\mu = \overline{\psi}\gamma^\mu\psi \quad \text{with } \partial_\mu j^\mu = 0} \tag{2.55}$$

as follows,

$$\partial_\mu j^\mu = \left(\partial_\mu\overline{\psi}\right)\gamma^\mu\psi + \overline{\psi}\gamma^\mu\left(\partial_\mu\psi\right) = \left(im\overline{\psi}\right)\psi + \overline{\psi}\left(-im\psi\right) = 0.$$

As shown later in Sec. 2.3, j^μ transforms as a 4-vector under Lorentz transformations. The associated charge

$$Q = \int d^3x\, j^0, \tag{2.56}$$

following from the charge density $j^0 = \overline{\psi}\gamma^0\psi = \psi^\dagger\psi$, is thus a Lorentz invariant and conserved quantity.

Before investigating the behaviour of spinors and composed quantities under Lorentz transformations, we study the solutions of the Dirac equation and the step towards a quantum field theory.

2.2.2 *Solutions of the Dirac equation*

Solutions of the Dirac equation can be found among the eigenfunctions of energy and momentum with eigenvalues p^0 and $\vec{p}$. Because of $H^2 = \vec{P}^2 + m^2$ any solution ψ of the Dirac equation has to fulfill the condition

$$\left(\Box + m^2\right)\psi = 0,$$

which means that each component of the spinor ψ is a solution of the Klein–Gordon equation. Momentum eigenfunctions correspond to the plane wave solutions

$$\psi \sim e^{\pm i\,px} \quad \text{with } p^0 = \sqrt{\vec{p}^2 + m^2}.$$

According to the sign, there are two classes of solutions, which will be treated separately.

2.2.2.1 *Solutions with e^{-ipx}*

Starting point is the ansatz with a 4-component spinor u for the amplitude,

$$\psi(x) = u(p)\,e^{-ipx} \quad \text{with } u(p) = \begin{pmatrix} u_1 \\ u_2 \\ u_3 \\ u_4 \end{pmatrix}. \tag{2.57}$$

With $\partial_\mu \psi = (-ip_\mu)\psi$ one finds that (2.57) is a solution of Eq. (2.50) if and only if $u(p)$ is a solution of the algebraic equation

$$\left(\gamma^\mu p_\mu - m\right) u(p) = 0$$

which in a compact notation reads

$$\boxed{\left(\not{p} - m\right) u(p) = 0}. \tag{2.58}$$

Therein, the notation (for an arbitrary 4-vector a^μ)

$$\not{a} = \gamma^\mu a_\mu \tag{2.59}$$

has been introduced. The algebraic Dirac equation (2.58) is a homogeneous system of linear equations with determinant $= 0$ und rank $= 2$, hence, there exist two linearly independent solutions. To find them, one expresses the 4-component spinor u in terms of two 2-component spinors φ and χ,

$$u = \begin{pmatrix} \varphi \\ \chi \end{pmatrix},$$

and obtains for Eq. (2.58) in the Dirac representation

$$\begin{pmatrix} (E - m)\mathbb{1} & -\vec{\sigma} \cdot \vec{p} \\ \vec{\sigma} \cdot \vec{p} & -(E + m)\mathbb{1} \end{pmatrix} \begin{pmatrix} \varphi \\ \chi \end{pmatrix} = \begin{pmatrix} 0 \\ 0 \end{pmatrix}. \tag{2.60}$$

This system corresponds to a single independent condition,

$$(E + m)\chi = \left(\vec{\sigma} \cdot \vec{p}\right)\varphi. \tag{2.61}$$

For φ any two linearly independent 2-component spinors can be chosen; the associated χ spinors are then determined according to

$$\chi = \frac{\vec{\sigma} \cdot \vec{p}}{E + m}\varphi. \tag{2.62}$$

The simplest choice is given by

$$\varphi = \begin{pmatrix} 1 \\ 0 \end{pmatrix}, \quad \begin{pmatrix} 0 \\ 1 \end{pmatrix}.$$

For physics reasons it may be useful to choose eigenstates of the helicity (2.38) with eigenvalues $\pm 1/2$, determined by $\varphi_\pm$ obeying

$$\left(\vec{\sigma} \cdot \vec{n}\right)\varphi_\pm = \pm\varphi_\pm, \quad \vec{n} = \vec{p}/|\vec{p}|, \tag{2.63}$$

to become

$$u_\pm(p) = \begin{pmatrix} \varphi_\pm \\ \frac{\vec{\sigma} \cdot \vec{p}}{E+m}\varphi_\pm \end{pmatrix} = \begin{pmatrix} \varphi_\pm \\ \pm\frac{|\vec{p}|}{E+m}\varphi_\pm \end{pmatrix}. \tag{2.64}$$

They fulfill the eigenvalue equation

$$\left(\vec{\Sigma}\cdot\vec{n}\right) u_\pm = \pm\, u_\pm \tag{2.65}$$

and are normalized according to

$$u_\pm^\dagger u_\pm = \left(\varphi_\pm^\dagger \varphi_\pm\right) \frac{2E}{E+M}. \tag{2.66}$$

The usual normalization is fixed by the convention (see also Eq. (2.80))

$$u_\pm^\dagger u_\pm = \overline{u}_\pm \gamma^0 u_\pm = 2E. \tag{2.67}$$

In the non-relativistic limit $|\vec{p}| \ll m$ the following simplification holds,

$$u_\pm(p) = \begin{pmatrix} \varphi_\pm \\ 0 \end{pmatrix} + \mathcal{O}\left(\frac{|\vec{p}|}{m}\right), \tag{2.68}$$

and in the ultra-relativistic case $|\vec{p}| \gg m$ one obtains

$$u_\pm(p) = \begin{pmatrix} \varphi_\pm \\ \pm\varphi_\pm \end{pmatrix} + \mathcal{O}\left(\frac{m}{|\vec{p}|}\right). \tag{2.69}$$

2.2.2.2 Solutions with e^{+ipx}

As in the previous case, starting point is the ansatz

$$\psi(x) = v(p)\, e^{ipx} \tag{2.70}$$

with a 4-component spinor $v(p)$. With $\partial_\mu \psi = (ip_\mu)\psi$ one finds that (2.70) is a solution of Eq. (2.50) if and only if $v(p)$ fulfills the algebraic equation

$$\left(\gamma^\mu p_\mu + m\right) v(p) = 0$$

or in compact notation

$$\boxed{\left(\not{p} + m\right) v(p) = 0}. \tag{2.71}$$

Also this homogeneous system of linear equations has the properties $\det = 0$ und rank $= 2$. Hence there exist two linearly independent solutions which can be found by building v from two 2-component spinors

$$v = \begin{pmatrix} \varphi \\ \chi \end{pmatrix}$$

and performing the same steps as done before. Insertion into Eq. (2.71) yields a single independent condition

$$(E + m)\,\varphi = (\vec{\sigma}\cdot\vec{p})\,\chi$$

which allows to choose any pair of linearly indendent 2-component spinors for χ; then the associated φ spinors follow from

$$\varphi = \frac{\vec{\sigma}\cdot\vec{p}}{E + m}\,\chi. \qquad (2.72)$$

For helicity states with helicity $\pm 1/2$ the v-spinors are given by

$$v_\pm(p) = \begin{pmatrix} \frac{\vec{\sigma}\cdot\vec{p}}{E+m}\,\chi_\mp \\ \chi_\mp \end{pmatrix} = \begin{pmatrix} \mp\frac{|\vec{p}|}{E+m}\,\chi_\mp \\ \chi_\mp \end{pmatrix} \qquad (2.73)$$

with $\chi_\pm$ obeying

$$(\vec{\sigma}\cdot\vec{n})\,\chi_\pm = \pm\chi_\pm, \quad \vec{n} = \vec{p}/|\vec{p}|. \qquad (2.74)$$

The spinors $v_\pm$ fulfill the eigenvalue equation[1]

$$(-\vec{\Sigma}\cdot\vec{n})\,v_\pm = \pm v_\pm. \qquad (2.75)$$

In the non-relativistic limit one obtains

$$v_\pm(p) = \begin{pmatrix} 0 \\ \chi_\mp \end{pmatrix} + \mathcal{O}\left(\frac{|\vec{p}|}{m}\right), \qquad (2.76)$$

and in the ultra-relativistic case

$$v_\pm(p) = \begin{pmatrix} \mp\chi_\mp \\ \chi_\mp \end{pmatrix} + \mathcal{O}\left(\frac{m}{|\vec{p}|}\right). \qquad (2.77)$$

Following the normalization of the u-spinors, the convention for the normalization of $v_\pm$ is chosen according to

$$v_\pm^\dagger v_\pm = \bar{v}_\pm \gamma^0 v_\pm = 2E. \qquad (2.78)$$

With the solutions for the u and v spinors one can specify a complete system of orthogonal functions as a basis for solutions of the Dirac equation, consisting of

$$\begin{aligned} u_\sigma(p)\,e^{-ipx} \quad &\text{with } (\not{p} - m)\,u_\sigma(p) = 0, \\ v_\sigma(p)\,e^{+ipx} \quad &\text{with } (\not{p} + m)\,v_\sigma(p) = 0, \end{aligned} \qquad (2.79)$$

where σ indicates the helicity eigenvalue (writing $\pm$ in simplification of $\pm 1/2$).

---

[1] For the extra minus sign see end of Sec. 2.2.3.

2.2.2.3 *Normalization*

The spinors u, v are normalized by the convention

$$\bar{u}_\sigma(p)\, u_{\sigma'}(p) = 2m\, \delta_{\sigma\sigma'}$$
$$\bar{v}_\sigma(p)\, v_{\sigma'}(p) = -2m\, \delta_{\sigma\sigma'}. \tag{2.80}$$

Note: When the normalization of φ and χ is chosen according to $\varphi^\dagger \varphi = \chi^\dagger \chi = 1$, an additional factor $\sqrt{E+m}$ is required for u and v in Eqs. (2.64) und (2.73) to be compatible with the convention (2.80).

2.2.2.4 *Polarization sum*

An important role, in particular for practical calculations in later course, is played by the polarization sums. Summing over the helicities of the spinors with the normalization (2.80) one obtains

$$\sum_\sigma u_\sigma(p)\, \bar{u}_\sigma(p) = \not{p} + m,$$
$$\sum_\sigma v_\sigma(p)\, \bar{v}_\sigma(p) = \not{p} - m. \tag{2.81}$$

Mathematically this corresponds to the completeness relation of u_σ and v_σ.

Note: The polarization sums with a different normalization,

$$\frac{\not{p} + m}{2m} = \mathcal{P}_+ \quad \text{and} \quad \frac{\not{p} - m}{2m} = \mathcal{P}_-,$$

are projection operators, as one can easily verify:

$$\mathcal{P}_+^2 = \mathcal{P}_+, \quad \mathcal{P}_-^2 = \mathcal{P}_-, \quad \mathcal{P}_+\mathcal{P}_- = \mathcal{P}_-\mathcal{P}_+ = 0.$$

They project on the 2-dimensional spaces spanned by u_σ and v_σ, which are orthogonal subspaces of the 4-dimensional spinor space.

2.2.2.5 *Physical interpretation of the solutions*

For the description of a particle with spin $1/2$ like the electron, one would expect that only two independent solutions are required corresponding to the two projections of the particle spin on a given axis, similar to the non-relativistic Pauli theory. The duplication of the number of solutions in the relativistic case is a new feature and necessitates an interpretation in terms of an additional degree of freedom. The attempt of an interpretation within relativistic quantum mechanics with a fixed particle number proves insufficient although a positive-definite density $\psi^\dagger\psi$ seems to be at hand. A satisfactory description, instead, needs the concept of quantum field theories.

Historically the solutions $\psi = v(p)e^{ipx}$ for "negative energy", according to $H\psi = -p^0\psi$, were considered problematic with respect to the physical understanding, whereas the solutions $u(p)e^{-ipx}$ for "positive energy" could be assigned to the electron. The question, however, why the electron is stable and does not fall into the lower-lying states with negative energy posed a riddle.

As a loophole, it was proposed by Dirac that all states with negative energy $-p^0 \leq -m$ are occupied, forming the "Dirac sea", with the consequence that according to the Pauli exclusion principle no electron with positive energy $p^0 \geq m$ can make a transition to a state of negative energy. By supply of an energy amount $\Delta E > 2m$, on the other hand, an electron from the sea can be transfered into a state of positive energy. Thereby a hole occurs in the sea which manifests itself as an electron with opposite charge and thus corresponds to an antiparticle. Although in this way the positron was predicted and particle–antiparticle creation represents a realistic physical process, the hole theory is highly questionable and only of historical interest. Conceptually it is a multi-particle theory with a variable number of particles and thus different from one-particle quantum mechanics.

The natural frame for such a multi-particle system is a quantum field theory. Accordingly, the Dirac equation is a field equation for a field operator $\psi(x)$ acting on the space of states for particles and antiparticles. In the concrete case of electrons and positrons one has the classification (for historical reasons)

$$
\begin{array}{rcl}
\text{particle} & = & \text{electron } e^- \\
\text{antiparticle} & = & \text{positron } e^+
\end{array}
$$

2.2.3 *Quantum field theoretical formulation*

Following the steps in Sec. 2.1 for the scalar field, particle and antiparticle states are combined in a common Hilbert space. The eigenstates of momentum, energy, and helicity with eigenvalues

$$
\vec{p}, \quad p^0 = \sqrt{\vec{p}^2 + m^2}, \quad \sigma = \pm\frac{1}{2}
$$

for either a particle or an antiparticle are described by ket vectors as follows (for concreteness the notation is chosen for the electron/positron system)

$$\text{particle state} \quad |e^-, p\sigma\rangle$$
$$\text{antiparticle state} \quad |e^+, p\sigma\rangle \tag{2.82}$$

Also here, there is a zero-particle state, the vacuum $|0\rangle$, with the normalization

$$\langle 0|0\rangle = 1. \tag{2.83}$$

The eigenstates of momentum and helicity are normalized according to

$$\langle e^\pm, p\sigma | e^\pm, p'\sigma'\rangle = 2p^0\,\delta_{\sigma\sigma'}\,\delta^3(\vec{p} - \vec{p}')$$
$$\langle e^\pm, p\sigma | e^\mp, p'\sigma'\rangle = 0. \tag{2.84}$$

The annihilation and creation operators for particles and antiparticles are defined by their mode of action with respect to

$$e^- \text{ annihilation} \quad c_\sigma(p)\,|e^-, p'\sigma'\rangle = 2p^0\,\delta_{\sigma\sigma'}\,\delta^3(\vec{p} - \vec{p}')\,|0\rangle$$
$$e^+ \text{ annihilation} \quad d_\sigma(p)\,|e^+, p'\sigma'\rangle = 2p^0\,\delta_{\sigma\sigma'}\,\delta^3(\vec{p} - \vec{p}')\,|0\rangle$$
$$c_\sigma(p)\,|e^+, p'\sigma'\rangle = d_\sigma(p)\,|e^-, p'\sigma'\rangle = 0$$
$$e^- \text{ creation} \quad c_\sigma^\dagger(p)\,|0\rangle = |e^-, p\sigma\rangle$$
$$e^+ \text{ creation} \quad d_\sigma^\dagger(p)\,|0\rangle = |e^+, p\sigma\rangle \tag{2.85}$$

Being fermionic operators, they fulfill canonical anti-commutation relations,

$$\{c_\sigma(p), c_{\sigma'}^\dagger(p')\} = 2p^0\,\delta_{\sigma\sigma'}\,\delta^3(\vec{p} - \vec{p}'),$$
$$\{d_\sigma(p), d_{\sigma'}^\dagger(p')\} = 2p^0\,\delta_{\sigma\sigma'}\,\delta^3(\vec{p} - \vec{p}'), \tag{2.86}$$

all other anti-commutators are zero.

Remark. The anti-commutation relations are the quantum field theoretical version of the Pauli exclusion principle. They imply $c_\sigma^\dagger(p)\,c_\sigma^\dagger(p) = d_\sigma^\dagger(p)\,d_\sigma^\dagger(p) = 0$, which means that two particles or antiparticles cannot occupy the same quantum state:

$$c_\sigma^\dagger(p)\,c_\sigma^\dagger(p)\,|0\rangle = 0, \quad d_\sigma^\dagger(p)\,d_\sigma^\dagger(p)\,|0\rangle = 0.$$

We can now write down the field operator $\psi(x)$ for the Dirac field as the general solution of the Dirac equation specified by a Fourier expansion in terms of the complete set of basis functions given in Eq. (2.79). Within the context of a quantum field theory, the coefficients of the expansion are the annihilation and creation operators for particles and antiparticles,

$$\psi(x) = \frac{1}{(2\pi)^{3/2}} \int \frac{d^3p}{2p^0} \sum_\sigma \left[c_\sigma(p)\, u_\sigma(p)\, e^{-ipx} + d_\sigma^\dagger(p)\, v_\sigma(p)\, e^{ipx} \right].$$

$$(2.87)$$

Wave functions of 1-particle states are now re-interpreted as matrix elements of the field operator between the vacuum and the 1-particle states,

$$\langle 0 \,|\, \psi(x) \,|\, e^-, p\sigma \rangle = \frac{1}{(2\pi)^{3/2}}\, u_\sigma(p)\, e^{-ipx},$$

$$\langle e^+, p\sigma \,|\, \psi(x) \,|\, 0 \rangle = \frac{1}{(2\pi)^{3/2}}\, v_\sigma(p)\, e^{ipx}.$$

$$(2.88)$$

They appear in the description of processes where particles are annihilated and/or antiparticles are created. Moreover, there are the adjoint wave functions,

$$\langle 0 \,|\, \overline{\psi}(x) \,|\, e^+, p\sigma \rangle = \frac{1}{(2\pi)^{3/2}}\, \overline{v}_\sigma(p)\, e^{-ipx},$$

$$\langle e^-, p\sigma \,|\, \overline{\psi}(x) \,|\, 0 \rangle = \frac{1}{(2\pi)^{3/2}}\, \overline{u}_\sigma(p)\, e^{ipx},$$

$$(2.89)$$

appearing in the description of processes where antiparticles are annihilated and/or particles are created.

2.2.3.1   Current operator and charge

Expressed in terms of the quantum field ψ the classical current (2.55) is converted into a current operator, and the integral over the 0-component becomes the charge operator

$$Q = \int d^3x \, j^0 = \int d^3x \, \psi^\dagger \psi.$$

$$(2.90)$$

Inserting the Fourier expansion (2.87) for ψ yields the representation

$$Q = \int \frac{d^3p}{2p^0} \sum_\sigma \left[c_\sigma^\dagger(p)\, c_\sigma(p) - d_\sigma^\dagger(p)\, d_\sigma(p) \right] \equiv \int \frac{d^3p}{2p^0} \left[N_{+\sigma}(p) - N_{-\sigma}(p) \right]$$

$$(2.91)$$

in terms of the number operators $N_{+\sigma}(p)$ and $N_{-\sigma}(p)$ for particles and antiparticles. The 1-particle states (2.82) are eigenstates of Q,

$$Q \, |e^{\mp}, p\sigma\rangle = \pm |e^{\mp}, p\sigma\rangle \tag{2.92}$$

with eigenvalues $+1$ for particles and -1 for antiparticles. For historical reasons, the charge of the electron is $+1$, in units of the elementary electron charge e.

2.2.3.2 *Mechanical observables*

For the physical system described by the Dirac equation the observables energy and momentum can be expressed in terms of the number operators as well,

$$H = \int d^3x \; \psi^{\dagger}(x) \big[-i\vec{\alpha} \cdot \nabla + \beta m \big] \psi(x)$$

$$= \int \frac{d^3p}{2p^0} \sum_{\sigma} p^0 \left[N_{+\sigma}(p) + N_{-\sigma}(p) \right], \tag{2.93}$$

$$\vec{P} = \int d^3x \; \psi^{\dagger}(x) \big[-i\,\nabla \big] \psi(x)$$

$$= \int \frac{d^3p}{2p^0} \sum_{\sigma} \vec{p} \left[N_{+\sigma}(p) + N_{-\sigma}(p) \right], \tag{2.94}$$

in analogy to Eq. (2.24) for the scalar field, but with an additonal helicity index. A deeper reason for the expressions (2.93) and (2.94) in terms of the spinor field can be found in Sec. 4.2 within Lagrangian field theory (from a physics point of view the representation in terms of number operators is evident). One can easily verify that the 1-particle states (2.82) are eigenstates of H and $\vec{P}$,

$$H \, |e^{\pm}, p\sigma\rangle = p^0 \, |e^{\pm}, p\sigma\rangle, \quad \vec{P} \, |e^{\pm}, p\sigma\rangle = \vec{p} \, |e^{\pm}, p\sigma\rangle. \tag{2.95}$$

An observable of particular importance for the Dirac field is the spin operator, represented in terms of the spinor field in the following way,

$$\vec{S} = \int d^3x \; \psi^{\dagger}(x) \left[\frac{1}{2}\vec{\Sigma} \right] \psi(x). \tag{2.96}$$

Although intuitively evident, a deeper reason for this expression is found in Lagrangian field theory, where one can show that $\vec{S}$ in Eq. (2.96) arises as the spin part of the conserved total angular momentum. For the interested reader the derivation is outlined in Sec. 4.2. For the helicities of particle and antiparticle states with momentum $\vec{p}$ the following relations hold, with $\vec{n} = \vec{p}/|\vec{p}|$,

$$(\vec{S}\cdot\vec{n})\left|e^-, p \pm \frac{1}{2}\right\rangle = \pm\frac{1}{2}\left|e^-, p \pm \frac{1}{2}\right\rangle$$

$$\Leftrightarrow (\vec{\Sigma}\cdot\vec{n})\, u_\pm(p) = \pm\, u_\pm(p) \tag{2.97}$$

$$(\vec{S}\cdot\vec{n})\left|e^+, p\pm, \frac{1}{2}\right\rangle = \pm\frac{1}{2}\left|e^+, p \pm \frac{1}{2}\right\rangle$$

$$\Leftrightarrow \left(-\vec{\Sigma}\cdot\vec{n}\right) v_\pm(p) = \pm\, v_\pm(p) \tag{2.98}$$

The minus sign in the last line originates from a necessary commutation of the fermionic operators $d \ldots d^\dagger$ in Eq. (2.96) when $\vec{S}$ acts on an antiparticle state. This explains the appearance of the reversed sign in the determination of the antiparticle spinors in Eq. (2.75).

2.2.3.3 *Charge conjugation*

Charge conjugation is an operation that transforms a particle into its antiparticle and vice versa. This operation, also denoted as C-conjugation,

$$C: \quad \text{particle} \quad \leftrightarrow \quad \text{antiparticle}$$

represents a symmetry transformation for free Dirac particles (and also for the electromagnetic interaction) and formally points out the basic ambiguity in the definition of a particle or antiparticle, respectively.

For the u and v spinors given in Eqs. (2.64) and (2.73) the following relations hold,

$$i\gamma^2\, u_\sigma^*(p) = v_\sigma(p),$$
$$i\gamma^2\, v_\sigma^*(p) = u_\sigma(p). \tag{2.99}$$

Thus, C-conjugation for the Dirac field can be defined in the following way,

$$C: \quad \psi \to \psi^C = i\gamma^2 \left(\psi^\dagger\right)^T \tag{2.100}$$

Applying C to the expansion (2.87) yields for the charge-conjugated Dirac field the expression

$$\psi^C(x) = \frac{1}{(2\pi)^{3/2}} \int \frac{d^3p}{2p^0} \sum_\sigma \left[c_\sigma^\dagger(p)\, v_\sigma(p)\, e^{ipx} + d_\sigma(p)\, u_\sigma(p)\, e^{-ipx} \right] \tag{2.101}$$

where particle and antiparticle have changed their role; now $d_\sigma^\dagger$ create *particles* and $c_\sigma^\dagger$ create *antiparticles*. The Dirac equation itself is invariant under the transformation (2.100); it is transformed into

$$\left(i\gamma^\mu \partial_\mu - m\right) \psi^C = 0 \tag{2.102}$$

as can be verified using the relation $\gamma^2 \, \gamma^{\mu\,*} \gamma^2 = \gamma^\mu$.

2.3 Lorentz Symmetry and Dirac Equation

Proving covariance of the Dirac equation requires a detailed discussion of the transformation properties of Dirac spinors under Lorentz transformations. We revert to Chap. 1 for notations and concepts.

2.3.1 *Lorentz transformations and spinors*

Consider a Lorentz transformation Λ of the space-time coordinates specified by the transformation matrix (Λ^μ_ν),

$$x'^\mu = \Lambda^\mu_\nu \, x^\nu.$$

In the quantum mechanical space of states of a spin-$\frac{1}{2}$ particle with mass m, this transformation is represented by an operator. The spin states in the rest frame can be defined by the eigenvalues $m_s = \pm\frac{1}{2}$ of S^3, the corresponding eigenvectors span the two-dimensional space of states. The elements of this space are called *spinors* and are denoted by φ_0.

2.3.1.1 *Rotation*

To start with a familiar matter, the description of rotations in quantum mechanics and their action on spin states is recalled. A rotation is determined by a rotation angle α and a rotation axis $\vec{n}_r$ with $|\vec{n}_r| = 1$. A 2-dimensional representation of the rotation is given by an operator D according to

$$\varphi_0 \to D\,\varphi_0 \quad \text{with } D = e^{i\alpha \vec{n}_r \cdot \vec{S}} = e^{i\frac{\alpha}{2}\,\vec{n}_r \cdot \vec{\sigma}} \tag{2.103}$$

where $\vec{\sigma} = (\sigma^1, \sigma^2, \sigma^3)$ denotes the triplet of Pauli matrices. The expansion of the exponential function and the properties of the Pauli matrices yield the following expression for the 2×2 matrix D,

$$D = \mathbb{1}\cos\frac{\alpha}{2} + i\left(\vec{n}_r \cdot \vec{\sigma}\right)\sin\frac{\alpha}{2}. \tag{2.104}$$

Under rotations, the Pauli matrices transform like a 3-dimensional vector,

$$D^\dagger \, \sigma^k \, D = R^k_l \, \sigma^l \tag{2.105}$$

with the elements R^k_l of the rotation matrix (1.17) within the 4-dimensional (Λ^μ_ν). For example, a rotation around the z-axis has the entries $\vec{n}_r \cdot \vec{\sigma} = \sigma^3$ and

$$(R^k_l) = \begin{pmatrix} \cos\alpha & \sin\alpha & 0 \\ -\sin\alpha & \cos\alpha & 0 \\ 0 & 0 & 1 \end{pmatrix}. \tag{2.106}$$

The verification of the relation (2.105) is recommended as an exercise.

2.3.1.2 *Boost*

Similar to a rotation, a boost is determined by an axis $\vec{n}_b$ with $|\vec{n}_b| = 1$ specifying the boost direction, and a further quantity in terms of the rapidity ϕ defined in Eq. (1.19). The boost matrix can thus be written in a way similar to a rotation matrix, with ϕ as an imaginary angle, and the 2-dimensional representation in spinor space has a corresponding structure,

$$\varphi_0 \to D\,\varphi_0 \tag{2.107}$$

with the 2×2 matrix

$$D = e^{\frac{\phi}{2}\,\vec{n}_b \cdot \vec{\sigma}} = \mathbb{1}\cosh\frac{\phi}{2} + (\vec{n}_b \cdot \vec{\sigma})\sinh\frac{\phi}{2}. \tag{2.108}$$

Example. For a boost with velocity v in the direction of the $-\vec{e}_1$ axis, the boost matrix is given by

$$(\Lambda^\mu_\nu) = \left(\begin{array}{cc|cc} \cosh\phi & \sinh\phi & 0 & 0 \\ \sinh\phi & \cosh\phi & 0 & 0 \\ \hline 0 & 0 & 1 & 0 \\ 0 & 0 & 0 & 1 \end{array}\right) \tag{2.109}$$

with the entries

$$\cosh\phi = \frac{1}{\sqrt{1-v^2}}, \quad \sinh\phi = \frac{v}{\sqrt{1-v^2}},$$

and the 2-dimensional representation reads as follows,

$$D = \mathbb{1}\cosh\frac{\phi}{2} - \sigma^1\sinh\frac{\phi}{2}. \tag{2.110}$$

2.3.1.3 *Boost and rotation*

Combining the previous two cases, a general Lorentz transformation Λ consisting of a boost and a rotation can be represented in the 2-dimensional spinor space by the following 2×2 matrix,

$$\Lambda \to D(\Lambda) = e^{\frac{1}{2}(i\alpha\,\vec{n}_r + \phi\,\vec{n}_b)\cdot\vec{\sigma}}. \tag{2.111}$$

Under a space inversion the axes transform differently,

$$\vec{n}_r \to \vec{n}_r, \quad \vec{n}_b \to -\vec{n}_b,$$

so that the representation (2.111) is modified according to

$$D(\Lambda) \to \overline{D}(\Lambda) = e^{\frac{1}{2}(i\alpha\,\vec{n}_r - \phi\,\vec{n}_b)\cdot\vec{\sigma}}. \tag{2.112}$$

The representation $\overline{D}$ is mathematically not equivalent to D; the correlation is given by

$$\overline{D} = T\,D^*\,T^{-1} \quad \text{with } T = i\sigma^2 = \begin{pmatrix} 0 & 1 \\ -1 & 0 \end{pmatrix}. \tag{2.113}$$

Accordingly, there exist two inequivalent representations of inversion-free Lorentz transformations Λ, namely

$$
\begin{aligned}
D: & \quad \varphi_0 \to D\,\varphi_0, \\
\overline{D}: & \quad \chi_0 \to \overline{D}\,\chi_0.
\end{aligned}
\tag{2.114}
$$

None of these representations is suitable to accommodate a space inversion **P** which exchanges D and $\overline{D}$. This means that representations of the full Lorentz symmetry require 4-dimensional matrices,

$$\Lambda: \quad \begin{pmatrix} \varphi_0 \\ \chi_0 \end{pmatrix} \longrightarrow \begin{pmatrix} D\,\varphi_0 \\ \overline{D}\,\chi_0 \end{pmatrix} = \begin{pmatrix} D & 0 \\ 0 & \overline{D} \end{pmatrix} \begin{pmatrix} \varphi_0 \\ \chi_0 \end{pmatrix} \tag{2.115}$$

$$\mathbf{P}: \quad \begin{pmatrix} \varphi_0 \\ \chi_0 \end{pmatrix} \longrightarrow \begin{pmatrix} \chi_0 \\ \varphi_0 \end{pmatrix} = \begin{pmatrix} 0 & \mathbb{1} \\ \mathbb{1} & 0 \end{pmatrix} \begin{pmatrix} \varphi_0 \\ \chi_0 \end{pmatrix} = \gamma^0 \begin{pmatrix} \varphi_0 \\ \chi_0 \end{pmatrix} \tag{2.116}$$

with the Dirac matrix γ^0 in the chiral representation, which appears as the natural one in this context. Hence, for this section the chiral representation of the Dirac matrices is used, listed in Eq. (2.48).

2.3.1.4 *Boost from the rest frame*

Now we can demonstrate how a boost transforming a particle at rest into a moving particle leads directly to the Dirac equation, which thus appears as a consequence of Lorentz symmetry. Consider a special boost from the rest frame of the particle to a moving inertial frame where the particle has a 4-momentum p^μ,

$$\text{rest frame } K \quad \longrightarrow \quad \text{moving frame } K'$$
$$\text{4-momentum } (m, \vec{0}) \longrightarrow (E, \vec{p}) = (p^0, \vec{p})$$

K' moves relatively to K into the direction opposite to the momentum, i.e. the boost axis $\vec{n}_b$ is given by

$$\vec{n}_b = -\vec{n}, \quad \vec{n} = \frac{\vec{p}}{|\vec{p}|}$$

and the rapidity is determined by

$$\cosh \phi = \frac{E}{m}.$$

The 2-dimensional representation matrices thus read as follows,

$$D = e^{-\frac{\phi}{2}\vec{n}\cdot\vec{\sigma}} = \mathbb{1}\cosh\frac{\phi}{2} - (\vec{n}\cdot\vec{\sigma})\sinh\frac{\phi}{2},$$

$$\overline{D} = e^{+\frac{\phi}{2}\vec{n}\cdot\vec{\sigma}} = \mathbb{1}\cosh\frac{\phi}{2} + (\vec{n}\cdot\vec{\sigma})\sinh\frac{\phi}{2}.$$

(2.117)

With the help of the formulae for hyperbolic functions

$$\cosh\frac{\phi}{2} = \sqrt{\frac{\cosh\phi + 1}{2}} = \sqrt{\frac{E+m}{2m}},$$

$$\sinh\frac{\phi}{2} = \sqrt{\frac{\cosh\phi - 1}{2}} = \sqrt{\frac{E-m}{2m}},$$

one obtains the explicit representations for boosting the 2-component spinors,

$$D\,\varphi_0 = \frac{E+m-\vec{p}\cdot\vec{\sigma}}{\sqrt{2m(E+m)}}\,\varphi_0 = \varphi(p) \equiv \varphi,$$

$$\overline{D}\,\chi_0 = \frac{E+m+\vec{p}\cdot\vec{\sigma}}{\sqrt{2m(E+m)}}\,\chi_0 = \chi(p) \equiv \chi,$$

(2.118)

and the 4-component spinors transform according to

$$\begin{pmatrix}\varphi_0 \\ \chi_0\end{pmatrix} \longrightarrow \begin{pmatrix}\varphi \\ \chi\end{pmatrix} = \begin{pmatrix}D\,\varphi_0 \\ \overline{D}\,\chi_0\end{pmatrix}.$$

(2.119)

Because of symmetry under a space inversion **P**, the particle states in the rest frame are eigenstates of **P** with eigenvalues ± 1. Accordingly, for the 4-component spinors in the rest frame one has the condition

$$\mathbf{P}\begin{pmatrix}\varphi_0 \\ \chi_0\end{pmatrix} = \begin{pmatrix}\chi_0 \\ \varphi_0\end{pmatrix} = \pm\begin{pmatrix}\varphi_0 \\ \chi_0\end{pmatrix} \Leftrightarrow \varphi_0 = \pm\chi_0.$$

The two cases are treated separately.

(i) $\boxed{\varphi_0 = +\chi_0}$

With this condition the momentum-dependent spinors φ and χ in Eq. (2.118) are given by

$$\varphi = \frac{E + m - \vec{p} \cdot \vec{\sigma}}{\sqrt{2m(E + m)}} \varphi_0,$$

$$\chi = \frac{E + m + \vec{p} \cdot \vec{\sigma}}{\sqrt{2m(E + m)}} \varphi_0. \qquad (2.120)$$

By use of the identity $(\vec{p} \cdot \vec{\sigma})^2 = \vec{p}^2$ and simple algebraic operations one can derive the following system of linear equations for φ and χ,

$$\left(E + \vec{p} \cdot \vec{\sigma}\right) \varphi = m \chi$$

$$\left(E - \vec{p} \cdot \vec{\sigma}\right) \chi = m \varphi \qquad (2.121)$$

written in matrix form as

$$\begin{pmatrix} E + \vec{p} \cdot \vec{\sigma} & -m\mathbb{1} \\ -m\mathbb{1} & E - \vec{p} \cdot \vec{\sigma} \end{pmatrix} \begin{pmatrix} \varphi \\ \chi \end{pmatrix} = \begin{pmatrix} 0 \\ 0 \end{pmatrix}.$$

Multiplying by γ^0 from left yields

$$\begin{pmatrix} -m\mathbb{1} & E - \vec{p} \cdot \vec{\sigma} \\ E + \vec{p} \cdot \vec{\sigma} & -m\mathbb{1} \end{pmatrix} \begin{pmatrix} \varphi \\ \chi \end{pmatrix} = \begin{pmatrix} 0 \\ 0 \end{pmatrix}$$

which can be written as follows,

$$\left[-m\mathbb{1} + E \underbrace{\begin{pmatrix} 0 & \mathbb{1} \\ \mathbb{1} & 0 \end{pmatrix} - \vec{p} \cdot \begin{pmatrix} 0 & \vec{\sigma} \\ -\vec{\sigma} & 0 \end{pmatrix}}_{E \gamma^0 - \vec{p} \cdot \vec{\gamma} = p_\mu \gamma^\mu = \not{p}} \right] \begin{pmatrix} \varphi \\ \chi \end{pmatrix} = \begin{pmatrix} 0 \\ 0 \end{pmatrix}$$

or in compact notation

$$\left(\not{p} - m\right) u = 0 \quad \text{with } u = \begin{pmatrix} \varphi \\ \chi \end{pmatrix}.$$

This is the Dirac equation in momentum space for the particle spinor u as given in Eq. (2.58), and $\psi(x) = u\, e^{-ipx}$ is a solution of the Dirac equation (2.50).

(ii) $\boxed{\varphi_0 = -\chi_0}$

In this case the system of equations analogue to (2.121) is given by

$$\left(E + \vec{p}\cdot\vec{\sigma}\right)\varphi = -m\,\chi$$
$$\left(E - \vec{p}\cdot\vec{\sigma}\right)\chi = -m\,\varphi. \tag{2.122}$$

With the same steps as in (i) one obtains

$$\left(-m\mathbf{1} - E\gamma^0 + \vec{p}\cdot\vec{\gamma}\right)\begin{pmatrix}\varphi\\\chi\end{pmatrix} = \left(-m\mathbb{1} - p_\mu\gamma^\mu\right)\begin{pmatrix}\varphi\\\chi\end{pmatrix} = \begin{pmatrix}0\\0\end{pmatrix}$$

or in compact notation

$$\left(\not{p} + m\right)v = 0 \quad \text{with } v = \begin{pmatrix}\varphi\\\chi\end{pmatrix}.$$

This is the Dirac equation in momentum space for the antiparticle spinor v as given in Eq. (2.71), and $\psi(x) = v\,e^{ipx}$ is a solution of the Dirac equation (2.50).

In summary, the spinors $u(p)$ and $v(p)$ can be obtained from the spinors in the rest frame by a boost into the frame with the particle momentum $p = (E, \vec{p})$,

$$u(p) = \begin{pmatrix}D\,\varphi_0\\\overline{D}\,\varphi_0\end{pmatrix}, \quad v(p) = \begin{pmatrix}D\,\chi_0\\-\overline{D}\,\chi_0\end{pmatrix} \tag{2.123}$$

with D und $\overline{D}$ given in Eq. (2.117). The choice $\varphi_0^\pm$ und $\chi_0^\pm$ with

$$\left(\vec{\sigma}\cdot\vec{n}\right)\varphi_0^\pm = \pm\varphi_0^\pm, \quad \left(\vec{\sigma}\cdot\vec{n}\right)\chi_0^\pm = \pm\chi_0^\pm$$

yields the spinors $u_\pm$ and $v_\pm$ refering to the states with helicity $\pm\frac{1}{2}$,

$$u_\pm(p) = \begin{pmatrix}D\,\varphi_0^\pm\\\overline{D}\,\varphi_0^\pm\end{pmatrix}, \quad v_\pm(p) = \begin{pmatrix}D\,\chi_0^\mp\\-\overline{D}\,\chi_0^\mp\end{pmatrix}. \tag{2.124}$$

2.3.2 *Covariance of the Dirac equation*

A Lorentz transformation Λ without inversion is represented in the 4-dimensional spinor space by the matrices

$$\Lambda \longrightarrow \begin{pmatrix}D & 0\\0 & \overline{D}\end{pmatrix} =: S(\Lambda) \tag{2.125}$$

in the chiral representation. Thus, a 4-component spinor $\psi(x)$ transforms according to

$$\psi(x) \longrightarrow \psi'(x') = S(\Lambda)\,\psi(x) \quad \text{with } x'^{\mu} = \Lambda^{\mu}_{\ \nu}\,x^{\nu}. \tag{2.126}$$

For the following discussion we introduce a compact notation for the Pauli and Dirac matrices by defining

$$\sigma^0 = \mathbb{1}, \quad \sigma^{\mu} = (\sigma^0, \vec{\sigma}), \quad \bar{\sigma}^{\mu} = (\sigma^0, -\vec{\sigma}). \tag{2.127}$$

With this notation the γ matrices in the chiral representation read as follows,

$$\gamma^{\mu} = \begin{pmatrix} 0 & \sigma^{\mu} \\ \bar{\sigma}^{\mu} & 0 \end{pmatrix}. \tag{2.128}$$

For the matrices D und $\overline{D}$ the following relations hold[2]

$$\boxed{D^{\dagger} = \overline{D}^{-1}, \quad \overline{D}^{\dagger} = D^{-1}} \tag{2.129}$$

as can easily be seen from the expressions (2.111) and (2.112).

This implies the fundamental properties

$$\overline{D}^{\dagger}\sigma^{\mu}\,\overline{D} = \Lambda^{\mu}_{\ \nu}\,\sigma^{\nu}$$
$$D^{\dagger}\bar{\sigma}^{\mu}\,D = \Lambda^{\mu}_{\ \nu}\,\bar{\sigma}^{\nu}. \tag{2.130}$$

In case of rotations these relations are reduced to those in Eq. (2.105). For boosts, it is sufficient to consider a boost along the $\vec{e}_1$ axis since the coordinate system can always be rotated accordingly. With the matrices given in Eqs. (2.109) and (2.110) one can easily verify the relations (2.130).

The transformation properties of σ^{μ} and $\bar{\sigma}^{\mu}$ are transfered to the 4×4 matrices $S(\Lambda)$ in Eq. (2.125) yielding the following important relation

$$\boxed{S^{-1}\gamma^{\mu}\,S = \Lambda^{\mu}_{\ \nu}\,\gamma^{\nu}}. \tag{2.131}$$

This can be checked by means of Eqs. (2.129) and (2.130) as follows,

$$\begin{pmatrix} \overline{D}^{\dagger} & 0 \\ 0 & D^{\dagger} \end{pmatrix} \begin{pmatrix} 0 & \sigma^{\mu} \\ \bar{\sigma}^{\mu} & 0 \end{pmatrix} \begin{pmatrix} D & 0 \\ 0 & \overline{D} \end{pmatrix}$$

$$= \begin{pmatrix} 0 & \overline{D}^{\dagger}\sigma^{\mu}\,\overline{D} \\ D^{\dagger}\bar{\sigma}^{\mu}\,D & 0 \end{pmatrix} = \begin{pmatrix} 0 & \Lambda^{\mu}_{\ \nu}\,\sigma^{\nu} \\ \Lambda^{\mu}_{\ \nu}\,\bar{\sigma}^{\nu} & 0 \end{pmatrix} = \Lambda^{\mu}_{\ \nu} \begin{pmatrix} 0 & \sigma^{\nu} \\ \bar{\sigma}^{\nu} & 0 \end{pmatrix}.$$

---

[2]For rotations one has $D = \overline{D}$ and the representation matrices are unitary.

The proof of covariance is now an easy exercise. Consider the Lorentz transformation (2.126) for the spinor ψ, in brief notation written as $\psi' = S\psi$. In the inertial frame of the primed quantities the Dirac equation reads

$$i\gamma^\mu \partial'_\mu \psi' - m\psi' = 0.$$

Since S commutes with ∂'_μ one can write

$$i\gamma^\mu S \partial'_\mu \psi - mS\psi = 0.$$

After multiplying with S^{-1} one obtains

$$i\left(S^{-1}\gamma^\mu S\right)\partial'_\mu \psi - m\psi = 0$$

and by means of Eq. (2.131),

$$i\gamma^\nu \Lambda^\mu_{\ \nu} \partial'_\mu \psi - m\psi = 0.$$

Finally, with the transformation (1.69) of the covariant 4-gradient one ends up with the Dirac equation in the frame of the unprimed quantities,

$$i\gamma^\nu \partial_\nu \psi - m\psi = 0.$$

All the steps made above are valid in both directions demonstrating thus covariance of the Dirac equation under Lorentz transforms that do not contain an inversion.

For completion also space inversions $\mathbf{P}$ have to be considered. As shown in Eq. (2.116), $\mathbf{P}$ is represented by $S(\mathbf{P}) = \gamma^0$ in the 4-dimensional spinor space, yielding

$$S^{-1}\gamma^0 S = \gamma^0, \quad S^{-1}\gamma^k S = -\gamma^k,$$

in accordance with the relation

$$S^{-1}\gamma^\mu S = \Lambda^\mu_{\ \nu}\gamma^\nu \quad \text{with } \left(\Lambda^\mu_{\ \nu}\right) = \begin{pmatrix} 1 & 0 & 0 & 0 \\ 0 & -1 & 0 & 0 \\ 0 & 0 & -1 & 0 \\ 0 & 0 & 0 & -1 \end{pmatrix}. \tag{2.132}$$

Hence, the steps made above for the inversion-free Lorentz transformations are valid as well implying covariance of the Dirac equation also for space inversions.

2.3.3 *Lorentz covariants*

By means of their transformation properties the following combinations of spinors and Dirac matrices can be classified as tensors of various ranks,

$$\overline{\psi}\,\psi \qquad \text{scalar}$$
$$\overline{\psi}\,\gamma_5\,\psi \qquad \text{pseudoscalar}$$
$$\overline{\psi}\,\gamma^\mu\,\psi \qquad \text{vector}$$
$$\overline{\psi}\,\gamma^\mu\gamma_5\,\psi \qquad \text{pseudovector, axialvector}$$
$$\overline{\psi}\,\sigma^{\mu\nu}\,\psi \qquad \text{2nd rank tensor}, \quad \sigma^{\mu\nu} = \tfrac{i}{2}\left[\gamma^\mu,\gamma^\nu\right].$$

The transformation properties can be immediately derived from $\psi \to S\psi$ and the relation (2.131) for the representation matrices S, together with

$$S^{-1} = \gamma^0\,S^\dagger\gamma^0. \qquad (2.133)$$

Two examples are made explicit, the others are left as exercises.

- Scalar

$$\overline{\psi'}\,\psi' = \left(S\psi\right)^\dagger\gamma^0 S\,\psi = \psi^\dagger S^\dagger\gamma^0 S\psi = \overline{\psi}\left(\gamma^0 S^\dagger\gamma^0\right)S\,\psi = \overline{\psi}S^{-1}S\psi = \overline{\psi}\,\psi$$

- Vector

$$\overline{\psi'}\gamma^\mu\psi' = \left(S\psi\right)^\dagger\gamma^0\gamma^\mu S\,\psi = \psi^\dagger S^\dagger\gamma^0\gamma^\mu S\psi = \overline{\psi}\left(\gamma^0 S^\dagger\gamma^0\right)\gamma^\mu S\,\psi$$
$$= \overline{\psi}\,S^{-1}\gamma^\mu S\,\psi = \Lambda^\mu_{\;\nu}\,\overline{\psi}\,\gamma^\nu\,\psi$$

For covariants involving γ_5 one has to distinguish between $S = \gamma^0$ for inversions and $S(\Lambda)$ for Lorentz transformations Λ without an inversion,

$$S^{-1}\gamma_5\,S = \begin{cases} +\gamma_5 & \text{for } S = S(\Lambda) \\ -\gamma_5 & \text{for } S = S(\mathbf{P}) \end{cases}$$

yielding different signs under space inversions for pseudoscalar and axialvector in comparison to scalar and vector.

2.4 Dirac Particle in an External Electromagnetic Field

Although relativistic one-particle quantum mechanics is not a consistent physics concept, it may be a useful tool for an approximative description of certain limiting cases. A typical example is a Dirac particle (electron or muon) at a low momentum scale in an external electromagnetic field which is treated as a classical field. The general case of an interaction with the electromagnetic quantum field corresponds to quantum electrodynamics and will be the content of the subsequent chapter.

Consider an electron (or muon) with mass m and charge e in a given electromagnetic field described by its classical 4-potential $(A^\mu) = (A^0, \vec{A})$. The *minimal substitution*[3]

$$P_\mu \rightarrow P_\mu - eA_\mu, \quad i\partial_\mu \rightarrow i\partial_\mu - eA_\mu \tag{2.134}$$

in the free Dirac equation (2.50) introduces the electromagnetic interaction between A_μ and the fermion described by the Dirac spinor ψ,

$$\gamma^\mu \left(i\partial_\mu - eA_\mu \right) \psi - m\psi = 0, \tag{2.135}$$

or equivalently expressed in terms of the $\vec{\alpha}$ and β matrices,

$$i\,\partial_0\psi = \left[\vec{\alpha}\cdot\left(-i\nabla - e\vec{A} \right) + eA_0 + \beta\,m \right]\psi. \tag{2.136}$$

For non-relativistic momenta this can be handled as a single-particle equation where ψ plays the role of a 1-particle wave function. Two important special cases are:

- electron in a static Coulomb field,
- electron in a static magnetic field.

Historically, the derived phenomena discussed below had a significant impact on the success of the Dirac equation.

2.4.1 *Electrostatic Coulomb field*

In case of an electron in the Coulomb potential

$$V(r) = e\,A_0(r) = -\frac{e^2}{4\pi r} \equiv -\frac{\alpha}{r}, \quad \vec{A} = 0,$$

Eq. (2.136) is the Dirac equation for the hydrogen atom. For stationary states with $\psi \sim e^{-iEt}$ one obtains the time-independent version

$$\left[-i\,\vec{\alpha}\cdot\nabla + \beta m + V(r) \right] = E\,\psi, \tag{2.137}$$

---

[3]In Sec. 1.4 the minimal substitution was introduced via Eq. (1.64) in the classical Hamiltonian to describe the interaction of a charged particle with an electromagnetic field.

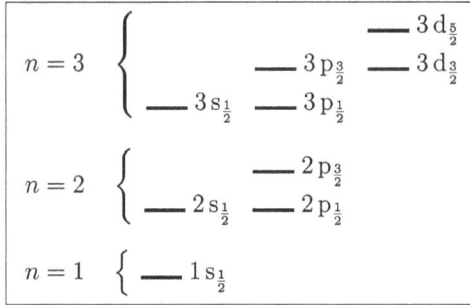

Fig. 2.1 Energy levels of the H-atom as predicted by the Dirac equation, with the common spectroscopic notation $s, p, d \ldots$ for $l = 0, 1, 2, \ldots$

as the eigenvalue equation for the energy levels. It can be solved exactly and predicts the fine structure of the spectrum of the H-atom, with spin–orbit interaction and relativistic corrections. The energy levels following from Eq. (2.137) are displayed schematically in Fig. 2.1. They are given by

$$E_{nj} = m \left(1 + \frac{\alpha^2}{(n - \epsilon_j)^2} \right)^{-1/2}$$

$$\text{with } \epsilon_j = j + \frac{1}{2} - \sqrt{\left(j + \frac{1}{2} \right)^2 - \alpha^2}$$

(2.138)

and are determined by the quantum numbers

$$n = 1, 2, \ldots \quad \text{and} \quad j = \frac{1}{2}, \frac{3}{2}, \ldots, n - \frac{1}{2}$$

corresponding to the main quantum number n and the total angular momentum j. The spectrum is independent of the orbital angular momentum l. Instead, for each j the two values $l = j \pm \frac{1}{2}$ determine the parity $(-1)^l$ of the wave function.

Not contained, however, is the *Lamb shift*, which is a typical quantum effect of the electromagnetic field and is thus a prediction of quantum electrodynamics (see Sec. 3.8).

2.4.2 *Static magnetic field*

A static magnetic field is described by the 4-potential

$$A^0 = 0, \quad \vec{A} = \vec{A}(\vec{x}).$$

(2.139)

In the non-relativistic approximation for $|\vec{p}| \ll m$, $|E - m| \ll m$ and with the ansatz in the Dirac representation

$$\psi = e^{-imt} \begin{pmatrix} \varphi \\ \chi \end{pmatrix} \quad \text{where } \chi = \mathcal{O}\left(\frac{|\vec{p}|}{m}\right) \cdot \varphi, \qquad (2.140)$$

an expansion in the leading terms yields a 2-component equation for φ,

$$i\,\partial_0 \varphi = \left[\frac{1}{2m} \left(-i\nabla - e\vec{A} \right)^2 - \frac{e}{2m}\, \vec{\sigma}\cdot\vec{B} \right] \varphi \qquad (2.141)$$

which is recognized as the Pauli equation. The spin part within the brackets,

$$\frac{e}{2m}\, \vec{\sigma}\cdot\vec{B} = \frac{e}{2m}\, g\, \vec{S}\cdot\vec{B},$$

describes the interaction of the electron magnetic moment with the magnetic field. The g-factor is predicted as $g = 2$; the small deviation from the experimental value, of the order 10^{-3}, is another quantum effect of the electromagnetic field.

The derivation of the Pauli equation in the non-relativistic limit is briefly explained. Starting from Eq. (2.136), the ansatz (2.140) leads to coupled equations for the 2-component spinors φ and χ,

$$i\,\partial_0 \varphi = \sigma\cdot\left(\vec{P} - e\vec{A}\right)\chi,$$

$$i\,\partial_0 \chi = \sigma\cdot\left(\vec{P} - e\vec{A}\right)\varphi - 2m\chi,$$

yielding χ as follows,

$$\chi = \frac{1}{2m}\sigma\cdot\left(\vec{P} - e\vec{A}\right)\varphi - \frac{1}{2m}\,i\partial_0\chi \approx \frac{1}{2m}\sigma\cdot\left(\vec{P} - e\vec{A}\right)\varphi$$

since the second term is of the order $\mathcal{O}(|\vec{p}|^2/m^2)$ and does not contribute to the expansion in the leading terms. Thus one obtains an equation for φ,

$$\begin{aligned} i\,\partial_0 \varphi &= \frac{1}{2m} \left[\sigma\cdot\left(\vec{P} - e\vec{A}\right) \right] \left[\sigma\cdot\left(\vec{P} - e\vec{A}\right) \right] \varphi \\ &= \frac{1}{2m} \left[\left(\vec{P} - e\vec{A}\right)^2 + i\vec{\sigma}\cdot\left(\vec{P} - e\vec{A}\right)\times\left(\vec{P} - e\vec{A}\right) \right] \varphi \qquad (2.142) \end{aligned}$$

by use of the identity

$$\left(\vec{\sigma}\cdot\vec{a}\right)\left(\vec{\sigma}\cdot\vec{b}\right) = \vec{a}\cdot\vec{b} + i\,\vec{\sigma}\cdot\left(\vec{a}\times\vec{b}\right)$$

which is valid for operators $\vec{a}$, $\vec{b}$ commuting with $\vec{\sigma}$. The further calculation obeying the commutation rules yields

$$\left(\vec{P} - e\vec{A}\right) \times \left(\vec{P} - e\vec{A}\right)\varphi = -e\vec{A} \times \vec{P}\varphi - e\vec{P} \times \left(\vec{A}\varphi\right)$$

$$= ie\vec{A} \times \nabla\varphi + ie\nabla \times \left(\vec{A}\varphi\right)$$

$$= ie\vec{A} \times \nabla\varphi + ie\left[\left(\nabla \times \vec{A}\right)\varphi - \vec{A} \times \nabla\varphi\right]$$

$$= ie\left(\nabla \times \vec{A}\right)\varphi.$$

Finally, by insertion into Eq. (2.142) the Pauli equation (2.141) comes about.

Chapter 3

Quantum Electrodynamics

Quantum electrodynamics (QED) is the fundamental theory of the electro-magnetic interaction. Historically QED was invented to describe the inter-action of electrons and positrons with photons. Today QED in a wider sense describes the interaction of photons with fundamental charged Dirac particles, such as $e^\pm$, $\mu^\pm$, $\tau^\pm$, and quarks.

3.1 Free Electromagnetic Field

The basis for the description of the classical electromagnetic field has already been given in Sec. 1.6. In the 4-dimensional formulation the vector potential $A^\mu(x)$ in Lorentz gauge $\partial_\mu A^\mu = 0$ fulfills Maxwell's equations in terms of an inhomogeneous wave equation

$$\Box A^\mu = j^\mu \tag{3.1}$$

with the source given by the 4-current j^μ. In QED the current is formed by the charged leptons and quarks.

As a first item we consider the free field with $j^\mu = 0$, which is a pure radiation field. By a gauge transformation, the solutions A^μ of the free wave equation can always be cast into the form

$$\left(A^\mu\right) = \left(0, \vec{A}\right) \quad \text{with } \partial_\mu A^\mu = \nabla \cdot \vec{A} = 0 \tag{3.2}$$

denoted as the *radiation gauge*. A complete set of solutions is given by the system of plane waves,

$$\epsilon_\lambda^\mu e^{\pm ikx} \tag{3.3}$$

with the wave vector $(k^\mu) = (k^0, \vec{k})$ complying with

$$k^0 = |\vec{k}|,$$

and the transverse polarization vectors

$$\left(\epsilon_\lambda^\mu\right) = \left(0, \vec{\epsilon}_\lambda\right), \quad \lambda = 1, 2, \tag{3.4}$$

which are orthogonal to the wave vector,

$$\epsilon_\lambda \cdot k = \epsilon_\lambda^\mu \cdot k_\mu = 0. \tag{3.5}$$

Because of $\nabla \cdot \vec{A} = 0$ the spatial components are orthogonal to $\vec{k}$; conventionally they are normalized and arranged as a right-handed trihedron,

$$\vec{\epsilon}_\lambda^{\,2} = 1, \quad \vec{\epsilon}_1 \cdot \vec{\epsilon}_2 = 0, \quad \vec{\epsilon}_1 \times \vec{\epsilon}_2 = \frac{\vec{k}}{|\vec{k}|}. \tag{3.6}$$

Equivalent to the linear polarisation vectors $\epsilon_{1,2}^\mu$ are the circular polarization vectors as complex linear combinations,

$$\epsilon_\pm^\mu = \frac{1}{\sqrt{2}}\left(\epsilon_1^\mu \pm i\epsilon_2^\mu\right), \tag{3.7}$$

which are orthogonal and normalized as well: $\epsilon_\lambda^* \cdot \epsilon_{\lambda'} = -\delta_{\lambda\lambda'}$ for $\lambda, \lambda' = \pm 1$.

In quantum field theory the circular polarizations correspond to the helicity states of the photon with helicity $\lambda = \pm 1$. Helicity $\lambda = 0$ does not exist; this would refer to a longitudinal polarization which is gauge dependent and thus unphysical.

The general solution of the free wave equation can now be written as a Fourier integral in terms of the plane waves (3.3) providing a complete system of orthogonal functions. The Fourier coefficients in this expansion are complex numbers in the classical case; for the quantized field they are the annihilation and creation operators of photons.

3.1.1   *Quantized electromagnetic field*

The quantized electromagnetic radiation field $A^\mu(x)$ represents the photon field, it is a field operator acting on the space of states of photons. Photons are particles with mass 0 and spin 1; the 4-momentum k^μ thus fulfills the energy–momentum relation of a massless particle,

$$(k^0)^2 - \vec{k}^{\,2} = 0, \quad k^0 = |\vec{k}|.$$

The space of states is constructed, in analogy to Klein–Gordon and Dirac particles, from the vacuum and the 1-particle states with momentum k^μ and helicity $\lambda = \pm 1$,

vacuum (0-photon state) $|0\rangle$

1-photon states $|k\lambda\rangle$

together with the two- and multi-particle states as symmetrized product states. The states are normalized according to

$$\langle 0 \,|\, 0 \rangle = 1, \quad \langle k\lambda \,|\, k'\lambda' \rangle = 2k^0 \,\delta_{\lambda\lambda'} \,\delta^3(\vec{k} - \vec{k}'). \tag{3.8}$$

The coefficients of the Fourier expansion of the field operator A^μ in terms of plane waves turn into the annihilation operators $a_\lambda(k)$ and creation operators $a_\lambda^\dagger(k)$ of photons with momentum k^μ und helicity λ,

$$A^\mu(x) = \frac{1}{(2\pi)^{3/2}} \int \frac{d^3k}{2k^0} \sum_\lambda \left[a_\lambda(k)\, \epsilon_\lambda^\mu \, e^{-ikx} + a_\lambda^\dagger(k)\, \epsilon_\lambda^{\mu\,*} \, e^{ikx} \right]. \tag{3.9}$$

Since for photons particles and antiparticles are the same, there is only one species of annihilation and creation operators acting on the photon states as follows,

$$a_\lambda^\dagger(k)\,|\,0\rangle = |k\lambda\rangle,$$
$$a_\lambda(k)|k'\lambda'\rangle = 2k^0 \,\delta^3(\vec{k} - \vec{k}')\,\delta_{\lambda\lambda'}\,|0\rangle. \tag{3.10}$$

Being bosonic operators, a_λ und $a_\lambda^\dagger$ fulfill canonical commutation rules like those for the scalar field, but with an additional spin index:

$$\left[a_\lambda(k), a_{\lambda'}^\dagger(k')\right] = 2k^0 \,\delta^3(\vec{k} - \vec{k}')\,\delta_{\lambda\lambda'}, \quad \left[a_\lambda(k), a_{\lambda'}(k')\right] = 0. \tag{3.11}$$

Wave functions of 1-photon states are the matrix elements

$$\langle 0 \,|\, A^\mu(x) \,|\, k\lambda \rangle = \frac{1}{(2\pi)^{3/2}}\, \epsilon_\lambda^\mu(k) e^{-ikx}, \quad \langle k\lambda \,|\, A^\mu(x) \,|\, 0 \rangle = \frac{1}{(2\pi)^{3/2}}\, \epsilon_\lambda^{\mu\,*}(k) e^{ikx}. \tag{3.12}$$

They are used in the description of processes where photons are annihilated (left) or created (right).

3.1.2 *Mechanical observables*

In correspondence to classical electrodynamics, Hamiltonian and momentum of the electromagnetic field are represented as integrals over the energy density and the momentum density in terms of the field operator,

$$H = \frac{1}{2} \int d^3x \left(\vec{E}^2 + \vec{B}^2 \right), \tag{3.13}$$

$$\vec{P} = \int d^3x \, \vec{E} \times \vec{B} \tag{3.14}$$

(see also Sec. 4.2.1). For the free radiation field in the radiation gauge (3.2) the field strengths are given by

$$\vec{E} = -\partial_0 \vec{A}, \quad \vec{B} = \nabla \times \vec{A}. \tag{3.15}$$

Inserting the Fourier expansion (3.9) for $\vec{A}$ yields the instructive representation

$$H = \int \frac{d^3k}{2k^0} \sum_\lambda k^0 \, N_\lambda(k),$$

$$\vec{P} = \int \frac{d^3k}{2k^0} \sum_\lambda \vec{k} \, N_\lambda(k) \tag{3.16}$$

in terms of the number operators $N_\lambda(k) = a_\lambda^\dagger(k) \, a_\lambda(k)$ for photons. Using the relations (3.10) it can easily be seen that the photon states $|p\sigma\rangle$ are eigenstates of H and $\vec{P}$,

$$H \, |p\sigma\rangle = p^0 \, |p\sigma\rangle, \quad \vec{P} \, |p\sigma\rangle = \vec{p} \, |p\sigma\rangle. \tag{3.17}$$

For photons as spin-1 particles the spin operator deserves an extra discussion. The spin operator for the radiation field is given by the expression

$$\vec{S} = \int d^3x \, (i\partial_0 A_k) \left(\vec{S}\right)^k_{\ l} A^l \tag{3.18}$$

with the matrices $\vec{S} = (\mathcal{S}^1, \mathcal{S}^2, \mathcal{S}^3)$ which are the generators of the 3-dimensional rotation group (see Sec. 4.2.2 for more details). For photons with momentum p the helicity is defined, like for Dirac particles, as the projection of the spin on the momentum direction,

$$\vec{S} \cdot \frac{\vec{p}}{|\vec{p}|} = \vec{S} \cdot \vec{n} = \int d^3x \, (i\partial_0 A_k) \left(\vec{S} \cdot \vec{n}\right)^k_{\ l} A^l. \tag{3.19}$$

Inserting the Fourier expansion (3.9) yields the representation

$$\vec{S} \cdot \vec{n} = \sum_{\lambda, \lambda'} \int \frac{d^3q}{2q^0} \left(-\epsilon_\lambda^*\right)_k \left(\vec{S} \cdot \vec{n}\right)^k_{\ l} \left(\epsilon_{\lambda'}\right)^l a_\lambda^\dagger(q) \, a_{\lambda'}(q) \tag{3.20}$$

in terms of the creation and annihilation operators and the circular polarization vectors ϵ_λ in Eq. (3.7) with $\lambda = \pm 1$.

With the help of the rules (3.10) and the orthogonality of the polarization vectors one finds: The photon state $|p\sigma\rangle$ with momentum p and helicity $\sigma = \pm 1$ fulfills the eigenvalue equation of the helicity operator

$$\left(\vec{S} \cdot \vec{n}\right) |p\sigma\rangle = \sum_\lambda \left(-\epsilon_\lambda^*\right)_k \left(\vec{S} \cdot \vec{n}\right)^k_{\ l} \left(\epsilon_\sigma\right)^l |p\sigma\rangle = \sum_\lambda \left(-\epsilon_\lambda^*\right)_k \left(\epsilon_\sigma\right)^k \sigma \, |p\sigma\rangle = \sigma \, |p\sigma\rangle$$

if the corresponding polarization vector ϵ_σ fulfills the condition

$$\left(\vec{S}\cdot\vec{n}\right)^k_{\ l}(\epsilon_\sigma)^l = \sigma\,(\epsilon_\sigma)^k \quad \text{or compact} \quad \left(\vec{S}\cdot\vec{n}\right)\vec{\epsilon}_\sigma = \sigma\,\vec{\epsilon}_\sigma, \tag{3.21}$$

which means that $\vec{\epsilon}_\sigma$ solves the eigenvalue problem of the matrix $\vec{S}\cdot\vec{n}$.

As an example we consider the case of a momentum in x^3 direction where $\vec{S}\cdot\vec{n} = S^3$. With the explict expression for S^3 (see Sec. 4.2.2) the condition (3.21) for the circular polarization vectors (3.7) with $\epsilon_\sigma = (0, \vec{\epsilon}_\sigma)$ can easily be verified:

$$S^3\,\vec{\epsilon}_\sigma = \begin{pmatrix} 0 & -i & 0 \\ i & 0 & 0 \\ 0 & 0 & 0 \end{pmatrix}\begin{pmatrix} 1 \\ \pm i \\ 0 \end{pmatrix}\frac{1}{\sqrt{2}} = (\pm 1)\begin{pmatrix} 1 \\ \pm i \\ 0 \end{pmatrix}\frac{1}{\sqrt{2}}.$$

3.2 Interacting Electromagnetic Field

Now we address the situation of the inhomogeneous wave equation (3.1) with the electromangetic current $j^\mu \sim \bar{\psi}\gamma^\mu\psi$, which is formed by the fermions $e^\pm$, $\mu^\pm$, ... with the respective Dirac fields, according to Eq. (2.55). From now on we make use of the notation

$$\Box A^\mu = e\,\bar{\psi}\gamma^\mu\psi \equiv e\,j^\mu \tag{3.22}$$

displaying the charge e as a coupling constant explicitly. The solution of this inhomogenous differential equation for a given boundary condition is found with the help of the appropriate Green function $D_{\mu\nu}$ as follows,

$$\boxed{A_\mu(x) = e\int d^4y\,D_{\mu\nu}(x-y)\,j^\nu(y)}. \tag{3.23}$$

$D_{\mu\nu}$ solves the inhomogeneous wave equation for a pointlike source,

$$\Box_{(x)}D_{\mu\nu}(x-y) = g_{\mu\nu}\,\delta^4(x-y). \tag{3.24}$$

Performing a Fourier transformation and using the formulae

$$\delta^4(x-y) = \int \frac{d^4Q}{(2\pi)^4}\,e^{-iQ(x-y)}, \tag{3.25}$$

$$D_{\mu\nu}(x-y) = \int \frac{d^4Q}{(2\pi)^4}\,e^{-iQ(x-y)}\,D_{\mu\nu}(Q) \tag{3.26}$$

one obtains an algebraic equation for the Fourier transformed $D_{\mu\nu}(Q)$,

$$-Q^2\,D_{\mu\nu}(Q) = g_{\mu\nu}. \tag{3.27}$$

The solution of this equation is written in the following way,

$$D_{\mu\nu}(Q) = \frac{-g_{\mu\nu}}{Q^2 + i\varepsilon}. \tag{3.28}$$

The appearance of an infinitesimal $\varepsilon > 0$ in the denominator is a regulation for integrating over the zeroes in the denominator of the Fourier integral for $D_{\mu\nu}(x - y)$ in Eq. (3.26). This regulation corresponds to a boundary condition for determining the Green function in space–time, yielding the *causal Green function*, also denoted as *Feynman propagator*.[1]

In case of the quantized field causal behaviour means: the propagator describes the propagation of a photon

$$\begin{array}{ll} \text{from } y \text{ to } x & \text{for } y^0 < x^0 \\ \text{from } x \text{ to } y & \text{for } x^0 < y^0 \end{array}$$

thus always from earlier to later times. The reason is found in the correlation between the causal Green function and the vacuum expectation value of a time-ordered product of field operators, the 2-point function,

$$i\, D_{\mu\nu}(x - y) = \langle 0|\, T A_\mu(x) A_\nu(y)\, |0\rangle, \tag{3.29}$$

where time ordering is defined by

$$T\, A_\mu(x) A_\nu(y) = \Theta(x^0 - y^0)\, A_\mu(x) A_\nu(y) + \Theta(y^0 - x^0)\, A_\nu(y) A_\mu(x). \tag{3.30}$$

For verifying the relation (3.29), one has to insert the Fourier representation (3.9). However, the summation exclusively over the transverse polarization vectors yields additional terms in the Fourier integral which can be written as follows,

$$-g_{\mu\nu} \rightarrow -g_{\mu\nu} - \frac{Q_\mu Q_\nu - (Q_\mu \eta_\nu + Q_\nu \eta_\mu)(\eta Q) + \eta_\mu \eta_\nu\, Q^2}{\vec{Q}^{\,2}} \tag{3.31}$$

with $(\eta_\mu) = (1, 0, 0, 0)$. This expression is the projector on the 2-dimensional subspace spanned by the transverse polarization vectors and thus orthogonal to (Q^μ) and $(0, \vec{Q})$. The terms with Q_μ and Q_ν are irrelevant for the calculation of physical quantities since photons couple to a conserved current, and the last term will be compensated by the Coulomb interaction. Hence, at the end only $g_{\mu\nu}$ contributes and Eq. (3.28) can be used

---

[1] Other regulations yield the retarded and advanced Green functions for solutions of the inhomogeneous wave equation as known in classical electrodynamics.

for practical calculations as the effective photon propagator (see also the discussion in Sec. 3.9).

With regard to later use in the Feynman graphs we introduce a graphical symbol for the photon propagator in momentum space, a wavy line carrying the momentum Q and two Lorentz indices,

$$i\,D_{\mu\nu}(Q) \qquad \overset{\text{\raisebox{2pt}{\wasysym}}}{\underset{\mu\quad Q\quad \nu}{}}$$

The factor i is by convention, according to the relation (3.29).

3.3 Interacting Dirac-Field

In Sec. 2.4 the interaction of charged Dirac particles with an external electromagnetic field described by a classical vector potential was addressed. Now the complete interaction with the quantized electromagnetic field $A^{\mu}(x)$ is taken into account. For a given fermion species, for example electron/positron, with the respective Dirac field $\psi(x)$, the Dirac equation including the electromagnetic interaction follows fom the free Dirac equation (2.50) by means of the minimal substitution $i\partial_{\mu} \to i\partial_{\mu} - eA_{\mu}$, yielding

$$\left(i\gamma^{\mu}\partial_{\mu} - m\right)\psi = e\,\gamma^{\mu}A_{\mu}\psi. \tag{3.32}$$

Treating the right-hand side as an inhomogeneity, a formal solution can specified by the method of Green functions,

$$\boxed{\psi(x) = e\int d^4y\, S(x-y)\,\gamma^{\mu}A_{\mu}(y)\,\psi(y)}. \tag{3.33}$$

Actually this formal "solution" is an integral equation for ψ; it is equivalent to the differential equation (3.32) together with a given boundary condition.

The Green function $S(x-y)$ of the Dirac equation is a 4×4-matrix, defined as a solution of the Dirac equation for a pointline inhomogeneity,

$$\left(i\gamma^{\mu}\partial_{\mu} - m\right) S(x-y) = \mathbf{1}\,\delta^4(x-y). \tag{3.34}$$

In analogy to the vector field one proceeds with the Fourier ansatz

$$S(x-y) = \int \frac{d^4Q}{(2\pi)^4}\, e^{-iQ(x-y)}\, S(Q) \tag{3.35}$$

converting Eq. (3.34) into an algebraic equation for $S(Q)$ in momentum space,

$$(\slashed{Q} - m)\, S(Q) = \mathbf{1} \tag{3.36}$$

with the solution

$$S(Q) = (\not{Q} - m)^{-1} = \frac{\not{Q} + m}{Q^2 - m^2 + i\varepsilon}. \tag{3.37}$$

Also here the $i\varepsilon$ regulation for integrating over the poles in the Fourier integral (3.35) corresponds to the boundary condition yielding the causal Green function or *Feynman propagator* for fermions. Since in the Dirac case particle and antiparticle are different, the causal Green function describes the propagation

$$
\begin{array}{llll}
\text{of the particle} & \text{from } y \text{ to } x & \text{for } y^0 < x^0 \\
\text{of the antiparticle} & \text{from } x \text{ to } y & \text{for } x^0 < y^0
\end{array}
$$

based on the correlation (written in terms of spinor components)

$$i\, S_{ab}(x - y) = \langle 0|\, T\, \psi_a(x)\, \overline{\psi}_b(y)\, |0\rangle \tag{3.38}$$

with the time ordering (the minus sign stems from the anticommuting fermion fields)

$$T\, \psi_a(x)\, \overline{\psi}_b(y) = \Theta(x^0 - y^0)\, \psi_a(x)\, \overline{\psi}_b(y) - \Theta(y^0 - x^0)\, \overline{\psi}_b(y)\, \psi_a(x). \tag{3.39}$$

For later use in the Feynman graphs another graphical symbol is introduced for the fermion propagator,

$$i\, S(Q) \qquad \underrightarrow{} \atop Q$$

carrying an arrow that indicates the direction of the charge flow of the *particle*; the associated momentum Q always points into the direction of the arrow.

3.4   Interaction and Time Evolution

In quantum mechanics as well as in quantum field theory there are two versions to describe the time evolution of a physical system.

- **Heisenberg picture**

  Observables are time dependent, $O(t)$, and their time evolution is determined by the Heisenberg equations of motion

$$i\, \frac{dO}{dt} = [O, H]. \tag{3.40}$$

• Schrödinger picture

States are time dependent, $|\psi(t)\rangle$, and their time evolution is determined by a unitary operator $U(t, t_0)$ according to

$$|\psi(t)\rangle = U(t, t_0) |\psi(t_0)\rangle \tag{3.41}$$

with U defined by the following differential equation and initial condition,

$$i\frac{dU}{dt} = H\, U, \quad U(t_0, t_0) = 1. \tag{3.42}$$

In both pictures, the Hamiltonian H is the fundamental dynamical quantity responsible for the time evolution of the system.

In QED the physical system consists of the electromagnetic field and (at least) one Dirac field. The system without interaction is described by the Hamiltonian $H_0 = H_0^{\text{Dirac}} + H_0^{\text{em}}$ for the Dirac field and the electromagnetic field; the interaction between both fields is determined by the interaction Hamiltonian

$$H_{\text{int}} = \int d^3x \; \mathcal{H}_{\text{int}}(x) \tag{3.43}$$

with a Hamiltonian density (conventionally denoted as Hamiltonian either)

$$\mathcal{H}_{\text{int}} = e\, j^\mu A_\mu \tag{3.44}$$

involving the current j^μ of the Dirac field (see also Sec. 3.9 and Sec. 4.4.5). The Hamiltonian of the entire system is thus given by

$$H = H_0 + H_{\text{int}}. \tag{3.45}$$

The equations of motion cannot be solved exactly. Approximate solutions are obtained by means of perturbation theory within the framework of the **interaction picture**, which is defined as follows.

- The time evolution of observables proceeds in the Heisenberg picture according to the free dynamics determined by H_0.
- The time evolution of states proceeds in the Schrödinger picture according to the dynamics determined by H_{int},

$$|\psi(t)\rangle = U(t, t_0) |\psi(t_0)\rangle \tag{3.46}$$

with the time evolution operator defined by

$$i\frac{dU}{dt} = H_{\text{int}}\, U \quad \text{and} \quad U(t_0, t_0) = 1. \tag{3.47}$$

Integration yields an integral equation which is equivalent to Eq. (3.47),

$$\boxed{U(t, t_0) - 1 = -i \int_{t_0}^{t} dt' \, H_{\text{int}}(t') \, U(t', t_0)}. \tag{3.48}$$

This equation can be solved iteratively and thus forms the basis for the perturbative treatment. By successive iteration one obtains the approximate solutions order by order for $n \geq 1$,

$$U^{(0)}(t, t_0) = 1,$$

$$U^{(n)}(t, t_0) = 1 - i \int_{t_0}^{t} dt' \, H_{\text{int}}(t') \, U^{(n-1)}(t', t_0). \tag{3.49}$$

In the following we restrict ourselves to the first approximation and make use of the simplified notation U instead of $U^{(1)}$. For $n = 1$ one obtains from Eq. (3.49)

$$U(t, t_0) = 1 - i \int_{t_0}^{t} dt' \, H_{\text{int}}(t') = 1 - i \int_{t_0}^{t} dt' \int d^3x \, \mathcal{H}_{\text{int}}(x). \tag{3.50}$$

The limit $t_0 \to -\infty$ and $t \to \infty$ yields the S-operator,

$$S \equiv \lim_{t_0 \to -\infty} \lim_{t \to \infty} U(t, t_0) = 1 - i \int d^4x \, \mathcal{H}_{\text{int}}(x). \tag{3.51}$$

The S-operator plays a fundamental role in scattering theory. The matrix elements $\langle f \,|\, S \,|\, i \rangle$ between asymptotically free particle states

$$|i\rangle \quad \text{initial state of free particles}$$
$$|f\rangle \quad \text{final state of free particles}$$

provide the transition amplitudes for scattering and decay processes. They are thus the probability amplitudes from which by squaring the transition probabilities are derived.

3.5   *S*-Matrix Elements and Feynman Graphs

For the matrix elements of the S-operator, the S-matrix elements, we introduce an abbreviated notation,

$$S_{fi} = \langle f \,|\, S \,|\, i \rangle. \tag{3.52}$$

The S-operator transforms the initial state $|i\rangle$ by time evolution into $S\,|i\rangle \equiv |i'\rangle$. The probability that the state $|f\rangle$ is contained in $|i'\rangle$ is given by

$$|\langle f | i' \rangle|^2 = |\langle f | S | i \rangle|^2 = |S_{fi}|^2.$$

The calculation of the matrix elements for given particle processes is thus of crucial importance for the prediction of reaction rates and cross sections. The first-order approximation (3.51) for S yields

$$S_{fi} = -i \int d^4 x \, \langle f | \mathcal{H}_{\text{int}}(x) | i \rangle \qquad (3.53)$$

under the assumption that initial and final states are not identical.

For the calculation of S-matrix elements a systematic method exists to display the matrix elements by Feynman graphs built by a set of Feynman rules. This would require, however, a somewhat extensive excursion into the formalism of quantum field theory and shall not be done within this introductory course. Instead, the procedure will be elucidated by means of a concrete example from which the Feynman rules can be read off that allow to obtain the matrix elements by graphical methods in an illustrative and efficient way.

Example: Electron–positron annihilation into muon pairs

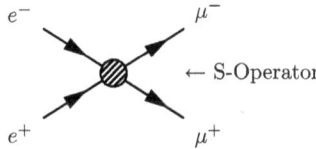

The specific example is the reaction $e^+ + e^- \to \mu^+ + \mu^-$. Initial and final states in the matrix element of the S-operator are defined by the momenta and helicities of the particles and antiparticles,

$$|i\rangle = |e^-, p\sigma_1\rangle \, |e^+, q\sigma_2\rangle$$
$$|f\rangle = |\mu^-, p'\sigma_1'\rangle \, |\mu^+, q'\sigma_2'\rangle$$

To simplify the notation, helicity indices will be suppressed in the following. Moreover, for distinguishing between the different types of fermions we choose a specific labeling of the various quantities according to

$\psi(x)$: $e^\pm$ field, $j^\nu = \overline{\psi}\gamma^\nu\psi$: $e^\pm$ current, u, v: $e^\pm$ spinors

$\Psi(x)$: $\mu^\pm$ field, $J^\nu = \overline{\Psi}\gamma^\nu\Psi$: $\mu^\pm$ current, U, V: $\mu^\pm$ spinors

For the calculation of S_{fi} we start from the expression (3.53) with $\mathcal{H}_{\text{int}} = J^\mu A_\mu$ and replace A_μ with the help of Eq. (3.23), according to the intuitive picture that the $\mu^{\pm}$ current couples to the vector potential that originates from the $e^{\pm}$ current,

$$S_{fi} = -ie \int d^4x \, \langle f | \, J^{\mu}(x) A_{\mu}(x) \, | i \rangle$$

$$= -ie^2 \int d^4x \, \langle f | \, \overline{\Psi}(x) \gamma^{\mu} \Psi(x) \int d^4y \, \underbrace{D_{\mu\nu}(x-y)} \, \overline{\psi}(y) \gamma^{\nu} \psi(y) \, | i \rangle$$

$$\int \frac{d^4Q}{(2\pi)^4} \, e^{-iQ(x-y)} \left(\frac{-g_{\mu\nu}}{Q^2 + i\varepsilon} \right) \qquad (3.54)$$

In the next step, for each field operator the respective Fourier expansion (2.87) is inserted. The resulting lengthy expression considerably shrinks by the action (2.85) of the annihilation operators on $| i \rangle$ and of the creation operators on $\langle f |$ (remember that creation operators act as annihilation operators on bra vectors to the left); from each expanded field operator only a single term remains, proportional either to $| 0 \rangle$ or to $\langle 0 |$ and with the respective particle/antiparticle wave function as coefficient. In summary, one finds

$$S_{fi} = -ie^2 \int \frac{d^4Q}{(2\pi)^4} \left(\frac{-g_{\mu\nu}}{Q^2 + i\varepsilon} \right) \overline{U}(p') \gamma^{\mu} V(q') \, \overline{v}(q) \gamma^{\nu} u(p) \, \langle 0|0 \rangle \left[\frac{1}{(2\pi)^{3/2}} \right]^4$$

$$\cdot \underbrace{\int d^4x \int d^4y \, e^{-iQ(x-y)} \, e^{i(p'+q')x} \, e^{-i(p+q)y}}$$

$$(2\pi)^8 \, \delta^4(p' + q' - Q) \, \delta^4(Q - p - q)$$

$$= (2\pi)^4 \, \delta^4(p' + q' - p - q) \left[\frac{1}{(2\pi)^{3/2}} \right]^4 \mathcal{M}_{fi}. \qquad (3.55)$$

The δ-functions correspond to momentum conservation at each endpoint of the propagator, hence they imply to insert

$$Q^2 = (p+q)^2 = (p'+q')^2. \qquad (3.56)$$

The appearance of the overall δ-function in the final result is universal, pointing out conservation of the total momentum $P_i = P_f$ in the scattering process. The genuine process-specific matrix element is the part of Eq. (3.55) labeled by $\mathcal{M}_{fi}$.

The process-specific matrix element can be written in the following way,

$$\mathcal{M}_{fi} = \overline{v}(q) i e \gamma^{\nu} u(p) \left(\frac{-i g_{\mu\nu}}{Q^2 + i\varepsilon} \right) \overline{U}(p') i e \gamma^{\mu} V(q') \qquad (3.57)$$

and has a graphical representation denoted as Feynman diagram or Feynman graph:

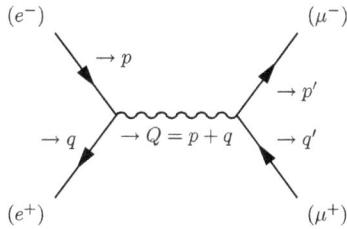

This Feynman graph contains three elements,

- the internal line corresponding to the photon propagator,
- the external lines corresponding to the wave functions of the initial and final state particles and antiparticles,
- the point connecting the lines, called vertex.

The Feynman graph is a specific combination of these elements following a prescription in terms of a set of rules, named *Feynman rules*. They translate the graphical symbols into well-defined analytical expressions. In the list below, the external lines are displayed with a dot to indicate the vertex where the line is connected to.

The matrix element $\mathcal{M}_{fi}$ is built according to the following set of Feynman rules.

	$\frac{-ig_{\mu\nu}}{Q^2+i\varepsilon}$	photon propagator
	$ie\gamma^\mu$	QED vertex
	$u(p)$	incoming particle
	$\bar{v}(p)$	incoming antiparticle
	$\bar{u}(p)$	outgoing particle
	$v(p)$	outgoing antiparticle

The arrow at the external fermion lines indicates the direction of the charge flow of the *particle*; hence, for antiparticles the arrow is oriented opposite to the momentum. Further rules are:

- 4-momentum conservation at each vertex,
- arrangement of the spinors *opposite* to the direction of the arrow.

The rules given above are not yet complete, since processes with external photons are not covered. The appearance of photons in the states $|i\rangle$ and/or $|f\rangle$ requires to extend the Feynman rules by including the photon wave functions and the fermion propagator. Our procedure is along the lines of muon pair production. Starting from Eq. (3.53) and substituting $\psi(x)$ by means of Eq. (3.33), one obtains the class of S-matrix elements describing processes with two photons and two fermions,

$$
\begin{aligned}
S_{fi} &= -ie \int d^4x \, \langle f \,|\, \overline{\psi}(x)\gamma^\mu \psi(x) \, A_\mu(x) \,|\, i \rangle \\
&= -ie^2 \int d^4x \int d^4y \, \langle f \,|\, \overline{\psi}(x)\gamma^\mu \, S(x-y) \, \gamma^\nu \psi(y) \, A_\nu(y) \, A_\mu(x) \,|\, i \rangle.
\end{aligned}
$$

$$(3.58)$$

Inserting the Fourier expansion for each A field generates the wave functions of initial and final state photons with momenta k and k', respectively,

$$
\langle f \,|\, \dots A_\mu(x) \dots \,|\, \gamma, k \dots \rangle = \cdots \, \epsilon_\mu(k) \, e^{-ikx},
$$

$$
\langle \gamma, k' \dots \,|\, \dots A_\mu(x) \dots \,|\, i \rangle = \cdots \, \epsilon_\mu^*(k') \, e^{ik'x},
$$

expressed in terms of the polarization vectors ϵ_μ (helicity indices are suppressed). The exponential factors become part of the δ-functions for momentum conservation after integration over x and y, hence the polarization vectors remain as external wave functions. In this way, the following additional Feynman rules are obtained which make the list complete:

$\xrightarrow{\quad}$ Q	$i\,\dfrac{\not{Q}+m}{Q^2-m^2+i\varepsilon}$	fermion propagator
$\sim\!\!\sim\!\!\bullet$ $k \to$	$\epsilon_\mu(k)$	incoming photon
$\bullet\!\!\sim\!\!\sim$ $k \to$	$\epsilon_\mu^*(k)$	outgoing photon

3.5.1   Modus operandi for the calculation of matrix elements

(1) Draw all connected graphs for given initial and final states.
(2) Assign the analytical expressions according to the Feynman rules.
(3) Sum all contributing graphs. Note: in case of exchanging two fermion lines a minus sign appears between the two graphs.

The result is $\mathcal{M}_{fi}$, which is related to S_{fi} in the following way,

$$S_{fi} = (2\pi)^4 \, \delta^4(P_f - P_i) \, \mathcal{M}_{fi} \left[\frac{1}{(2\pi)^{3/2}}\right]^n$$

$n = n_i + n_f :$ number of particles in $|i\rangle$ and $|f\rangle$

$P_i :$ total momentum in $|i\rangle$

$P_f :$ total momentum in $|f\rangle$

The application of the method is illustrated by examples of typical QED processes. The $i\varepsilon$ term in the denominators of the propagators can be dropped because of $Q^2 \neq 0$ in these cases (as well as in case of Eq. (3.57) for muon pair production).

- **Compton scattering:** $e^- + \gamma \to e^- + \gamma$

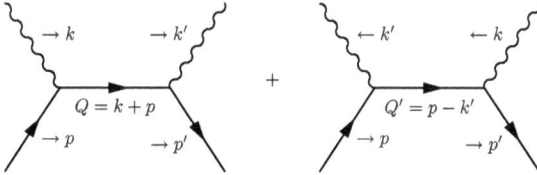

$$\mathcal{M} = \bar{u}(p')ie\gamma^\mu \, i\frac{\slashed{p} + \slashed{k} + m_e}{(p+k)^2 - m_e^2} \, ie\gamma^\nu u(p) \, \epsilon_\mu^*(k') \, \epsilon_\nu(k)$$

$$+ \bar{u}(p')ie\gamma^\nu \, i\frac{\slashed{p} - \slashed{k}' + m_e}{(p-k')^2 - m_e^2} \, ie\gamma^\mu u(p) \, \epsilon_\mu^*(k') \, \epsilon_\nu(k).$$

- **Electron–positron annihilation into photons:** $e^- + e^+ \to \gamma + \gamma$

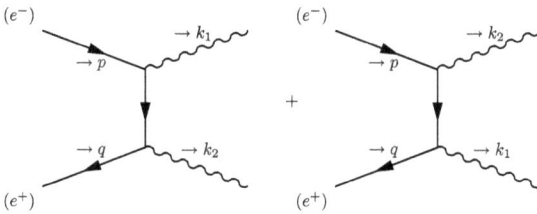

$$\mathcal{M} = \bar{v}(q)ie\gamma^\mu \; i\frac{\not{p}- \not{k}_1 + m_e}{(p-k_1)^2 - m_e^2} \; ie\gamma^\nu u(p) \; \epsilon^*_\nu(k_1) \, \epsilon^*_\mu(k_2)$$

$$+ \bar{v}(q)ie\gamma^\mu \; i\frac{\not{p}- \not{k}_2 + m_e}{(p-k_2)^2 - m_e^2} \; ie\gamma^\nu u(p) \; \epsilon^*_\nu(k_2) \, \epsilon^*_\mu(k_1).$$

- **Electron–electron scattering:** $e^- + e^- \to e^- + e^-$

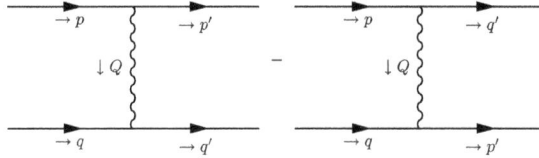

$$\mathcal{M} = \bar{u}(p')ie\gamma^\mu u(p) \frac{(-ig_{\mu\nu})}{(p-p')^2} \bar{u}(q')ie\gamma^\nu u(q)$$

$$- \bar{u}(q')ie\gamma^\mu u(p) \frac{(-ig_{\mu\nu})}{(p-q')^2} \bar{u}(p')ie\gamma^\nu u(q).$$

The relative minus sign between the diagrams results from the exchange of the fermion lines refering to p' and q'. Since the two electrons cannot be distinguished, both options for assigning the momentum p' or q' to the external line at either vertex have to be taken into account. The minus sign is a consequence of the anticommuting fermion fields and thus of the Pauli principle.

3.5.2   *Extension to all fermions*

So far only leptons have been considered, which carry the charge e, the negative elementary charge of the electron. From now on a new convention is agreed on:

$$e > 0 \quad \text{with} \quad \frac{e^2}{4\pi} = \alpha = \frac{1}{137.036\ldots}. \tag{3.59}$$

The numerical value of e is thus determined by the electromagnetic fine structure constant α, and the physical charge of the electron is defined as $e\,Q_e$ with $Q_e = -1$.

Leptons and quarks form the fundamental fermions. Their charges in units of e are defined as follows

e, μ, τ leptons:	$Q_{e,\mu,\tau}$	$= -1$
u, c, t quarks:	$Q_{u,c,t}$	$= +2/3$
d, s, b quarks:	$Q_{d,s,b}$	$= -1/3$

Each fermion species f is described by a Dirac field ψ_f, and the repective electromagnetic current is given by

$$j_f^\mu = Q_f \, \overline{\psi}_f \gamma^\mu \psi_f. \tag{3.60}$$

Accordingly, the vertex in the Feynman rules has to be replaced by

$$ieQ_f \gamma^\mu, \quad f = e, \mu, \tau, u, d, \ldots$$

and is thus applicable for all fermions. All the other QED Feynman rules remain unchanged.

3.6 Cross Section

The calculation of S-matrix elements provides the probability amplitudes for physical scattering processes. In order to get to the observables some more steps are required. The approach described here

- is valid for arbitrary reactions $a + b \to a_1 + a_2 + \cdots + a_n$ comprising also particle annihilation and creation,
- contains full relativistic kinematics and is Lorentz invariant,
- can be applied to all processes independent of the underlying interaction, although the application here is primarily to QED.

Starting point is the S-matrix element

$$S_{fi} = (2\pi)^4 \, \delta^4(P_f - P_i) \, \mathcal{M}_{fi} \left[\frac{1}{(2\pi)^{3/2}} \right]^{2+n} \equiv (2\pi)^4 \, \delta^4(P_f - P_i) \, \mathcal{T}_{fi}. \tag{3.61}$$

Squaring leads to the probability for the transition from $|i\rangle$ to $|f\rangle$. One has to take into account, however, that the final states are part of the continuous spectrum, as free particle states with momenta $p_1, \ldots p_n$, and thus only a differential transition probability can be defined,

$$dW = |S_{fi}|^2 \, d\Phi, \tag{3.62}$$

for the transition into the phase space element

$$d\Phi = \frac{d^3 p_1}{2p_1^0} \cdots \frac{d^3 p_n}{2p_n^0} \tag{3.63}$$

where $|S_{fi}|^2$ now plays the role of the probability density.

Here we encounter a problem: the square of the δ-function is mathematically not defined. The reason for the δ-function is the continuous spectrum which by itself is a consequence of an infinitely extended space-time. Therefore, to circumvent the problem, it is advisable to utilize a finite 4-dimensional volume, instead, for regularisation. This means a replacement

$$\int_{-\infty}^{\infty} dx^i \rightarrow \int_{-L/2}^{L/2} dx^i, \qquad \int_{-\infty}^{\infty} dt \rightarrow \int_{-T/2}^{T/2} dt \tag{3.64}$$

yielding the approximation, in the limit of large T,

$$(2\pi)\,\delta(\omega) \approx \int_{-T/2}^{T/2} dt\, e^{i\omega t} = \frac{2}{\omega} \sin\left(\frac{\omega T}{2}\right)$$

and accordingly

$$[(2\pi)\,\delta(\omega)]^2 \approx T \cdot \underbrace{\frac{2}{\omega T} \sin\left(\frac{\omega T}{2}\right)}_{=\,1 \text{ for } \omega = 0} \cdot \underbrace{\frac{2}{\omega} \sin\left(\frac{\omega T}{2}\right)}_{\approx\, 2\pi\,\delta(\omega)} \approx T \cdot 2\pi\,\delta(\omega).$$

Analogously one obtains for each space direction in the limit of large L,

$$[(2\pi)\,\delta(k)]^2 \approx L \cdot 2\pi\,\delta(k)$$

and consequently, with $V = L^3$, one can replace

$$[(2\pi)^4\,\delta^4(P_f - P_i)]^2 \rightarrow T \cdot V \cdot (2\pi)^4\,\delta^4(P_f - P_i). \tag{3.65}$$

The differential transition probability (3.62) thus yields the differential transition rate

$$dw = \frac{dW}{T} = V \cdot (2\pi)^4\,\delta^4(P_f - P_i)\, d\Phi\, |T_{fi}|^2 \tag{3.66}$$

and by normalization the *differential cross section* in the laboratory frame (target at rest),

$$d\sigma = \frac{dw}{|\vec{j}_a| N_b}.$$ (3.67)

The flux factor in the denominator contains the quantitites

$\vec{j}_a$ incoming particle-current density,

N_b number of target particles,

owing to the convention

$p_a = (p_a^0, \vec{p}_a)$ incoming momentum,

$p_b = (m_b, \vec{0})$ target momentum (rest frame).

For the determination of particle number and current density, the continuum normalization of the momentum states has to be adapted to the finite volume V with a discrete spectrum,

$$\langle p|p' \rangle = 2p^0 \, \delta^3(\vec{p} - \vec{p}') \rightarrow 2p^0 \, \frac{V}{(2\pi)^3} \, \delta_{\vec{p}\vec{p}'}$$ (3.68)

as one can see from the replacement of the δ-function by means of Eq. (3.64). Now the normalization is finite, however, different from unity,

$$\langle p \,|\, p \rangle = 2p^0 \, \frac{V}{(2\pi)^3} = N,$$ (3.69)

which means that N particles are contained in the volume V. Accordingly, the particle density is given by

$$\rho = \frac{N}{V} = \frac{2p^0}{(2\pi)^3}.$$ (3.70)

Multiplication with the velocity yields the particle current density $\vec{j} = \rho \vec{v}$. With reference to the cross section the follwing quantites are needed,

$$\vec{j}_a = \rho_a \vec{v}_a = \rho_a \frac{\vec{p}_a}{p_a^0} = \frac{\vec{p}_a}{(2\pi)^3},$$

$$N_b = \frac{2m_b}{(2\pi)^3} V.$$

Insertion into Eq. (3.67) leads to

$$d\sigma = \frac{(2\pi)^6}{2m_b \cdot 2|\vec{p}_a| \cdot V} \, dw$$

and finally, with dw from Eq. (3.66), to the result

$$d\sigma = \frac{(2\pi)^{10}}{2|\vec{p}_a| \cdot 2m_b} |\mathcal{T}_{fi}|^2 \, \delta^4(P_f - P_i) \, d\Phi. \qquad (3.71)$$

The momenta refer to the laboratory frame, the flux factor, however, can be written in a Lorentz-invariant way by means of

$$|\vec{p}_a| \, m_b = \sqrt{(p_a \cdot p_b)^2 - m_a^2 \, m_b^2},$$

yielding the manifestly invariant cross section

$$\boxed{d\sigma = \frac{(2\pi)^{10}}{4\sqrt{(p_a \cdot p_b)^2 - m_a^2 m_b^2}} |\mathcal{T}_{fi}|^2 \, \delta^4(P_f - P_i) \, d\Phi} \qquad (3.72)$$

The invariant form of the cross section has the advantage that it can be evaluated in each reference frame, among others in the center-of-mass system (CMS) which is defined by $\vec{p}_a + \vec{p}_b = 0$.

3.6.1 *Special case: 2-particle processes*

An important special case is the cross section for $2 \to 2$ reactions, with momenta

$$P_i = p_a + p_b, \quad P_f = p_1 + p_2.$$

Replacing $\mathcal{T}_{fi}$ by $\mathcal{M}_{fi}$ yields the general 2-particle cross section

$$d\sigma = \frac{(2\pi)^{-2}}{4\sqrt{(p_a \cdot p_b)^2 - m_a^2 m_b^2}} |\mathcal{M}_{fi}|^2 \, \delta^4(P_f - P_i) \, \frac{d^3 p_1}{2p_1^0} \frac{d^3 p_2}{2p_2^0}. \qquad (3.73)$$

For practical applications, in particular the CMS and the laboratory frame are of importance. Hence, we give concrete formulae for these cases including kinematical details.

Center-of-mass frame. In the CMS with $\vec{P}_i = \vec{P}_f = 0$ the momenta of the incoming and outgoing particles are given by

$$\vec{p}_a = -\vec{p}_b \equiv \vec{p}, \quad \vec{p}_1 = -\vec{p}_2 \equiv \vec{p}\,'.$$

The scattering angle θ is defined as the angle between $\vec{p}_1$ und $\vec{p}_a$. Choosing the $\vec{p}_a$ direction as the polar axis one can express the volume element $d^3 p_1$

in polar coordinates as follows,

$$d^3 p_1 = |\vec{p}_1|^2 \, d|\vec{p}_1| \, d\Omega, \quad d\Omega = d\cos\theta \, d\varphi, \tag{3.74}$$

with θ und φ as polar angle and azimuth. After integration over $d|\vec{p}_1|$ and $d^3 p_2$ with the help of the δ-function the differential cross section reads

$$d\sigma = \frac{1}{64\pi^2 s} \frac{|\vec{p}'|}{|\vec{p}|} |\mathcal{M}_{fi}|^2 \, d\Omega \tag{3.75}$$

with the invariant variable s representing the square of the total energy in the CMS,

$$s = (p_a + p_b)^2 = (p_a^0 + p_b^0)^2 = E_{\text{CMS}}^2. \tag{3.76}$$

Momentum conservation yields the relations

$$|\vec{p}| = \frac{1}{2\sqrt{s}} \sqrt{(s - m_a^2 - m_b^2)^2 - 4m_a^2 m_b^2},$$
$$|\vec{p}'| = \frac{1}{2\sqrt{s}} \sqrt{(s - m_1^2 - m_2^2)^2 - 4m_1^2 m_2^2}. \tag{3.77}$$

In the high-energy approximation where masses are negligible these expressions simplify to

$$|\vec{p}| = |\vec{p}'| = p^0 = p'^{\,0} = \frac{\sqrt{s}}{2}. \tag{3.78}$$

Laboratory frame. The laboratory frame is defined as the frame with the target at rest, $p_b = (m_b, \vec{0})$. The scattering angle θ denotes the angle between $\vec{p}_1$ and $\vec{p}_a$. Choosing $\vec{p}_a$ as the direction of the polar axis, $d^3 p_1$ can be expressed in polar coordinates as done in Eq. (3.74). After integration over $d^3 p_2$ and $d|\vec{p}_1|$ by means of the δ-function one obtains the following form of the differential cross section,

$$d\sigma = \frac{1}{64\pi^2 m_b^2} \left(\frac{p_1^0}{p_a^0}\right)^2 |\mathcal{M}_{fi}|^2 \, d\Omega \tag{3.79}$$

with the relation

$$\frac{p_1^0}{p_a^0} = \left(1 + \frac{p_a^0}{m_b}(1 - \cos\theta)\right)^{-1} \tag{3.80}$$

as a consequence of momentum conservation. For $m_a = m_1 = 0$ this form of the cross section is exact, otherwise it holds in the high-energy approximation with $|\vec{p}_a| \gg m_a$, $|\vec{p}_1| \gg m_1$. Moreover, it is also valid for $|\vec{p}_a| \ll m_b$ when the recoil is negligible and $p_1^0 \approx p_a^0$.

3.6.2 Unpolarized cross section

In the S-matrix elements the particle states of $|i\rangle$ and $|f\rangle$ are determined by the momenta $p_a, p_b, p_1, \ldots p_n$ and the helicities $\sigma_a, \sigma_b, \sigma_1, \ldots \sigma_n$, with the consequence that the matrix elements are helicity amplitudes,

$$\mathcal{M}_{fi} = \mathcal{M}_{fi}(\sigma_a, \sigma_b, \sigma_1, \ldots \sigma_n),$$

and the resulting cross sections refer to polarized particles with definite helicities. Normally, however, final state helicities are not measured and initial state particles are unpolarized, i.e. all helicities are of equal weight. In order to allow for these aspects and to obtain the respective unpolarized cross sections one has

to sum over $\sigma_1, \ldots \sigma_n$,
to average over σ_a, σ_b.

The corresponding instruction reads: replace in the formula for the cross section

$$|\mathcal{M}_{fi}|^2 \to \overline{|\mathcal{M}_{fi}|^2} = \frac{1}{N_a N_b} \sum_{\sigma_a, \sigma_b} \sum_{\sigma_1, \ldots \sigma_n} |\mathcal{M}_{fi}|^2 \qquad (3.81)$$

with N_a and N_b denoting the number of helicities of the particles a and b. In doing so, one has to take account of

$$N = 2s + 1 \quad \text{for spin } s \text{ and mass} \neq 0,$$
$$N = 2 \qquad \text{for spin } s \text{ and mass} = 0.$$

In case of massless particles with spin s only the two maximum helicities $\lambda = \pm s$ are realized in physical states. For photons these are the helicities $\lambda = \pm 1$ going with the two circular polarization vectors.

As examples we consider the processes of fermion pair production in electron–positron annihilation and Compton scattering. Summing and averaging over the helicities has to be performed according to

$$e^+ e^- \to f \bar{f} : \quad \overline{|\mathcal{M}_{fi}|^2} = \frac{1}{2} \cdot \frac{1}{2} \sum_{\sigma_a, \sigma_b} \sum_{\sigma_1, \sigma_2} |\mathcal{M}_{fi}|^2,$$

$$\gamma e^- \to \gamma e^- : \quad \overline{|\mathcal{M}_{fi}|^2} = \frac{1}{2} \cdot \frac{1}{2} \sum_{\lambda, \sigma} \sum_{\lambda', \sigma'} |\mathcal{M}_{fi}|^2.$$

3.6.3 *Decay width*

With regard to later applications the definition of the decay width for an unstable particle is introduced. Consider a particle with mass M which can decay according to $a \rightarrow a_1 + \cdots + a_n$ into n other particles with masses $m_1, \ldots, m_n$ and momenta $p_1, \ldots, p_n$,

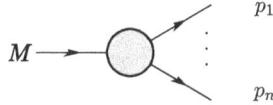

with the total momentum $P_f = p_1 + \cdots + p_n$. In the rest frame of the decaying particle the momentum is given by

$$P_i = p_a = (M, \vec{0}).$$

The differential decay width $d\Gamma$ is defined as the transition rate into the phase space element $d\Phi$, normalized to the number N_a of particles in the volume V,

$$N_a = \frac{2p_a^0}{(2\pi)^3} V = \frac{2M}{(2\pi)^3} V.$$

With dw in Eq. (3.66) and $d\Phi$ in Eq. (3.63) one obtains

$$d\Gamma = \frac{dw}{N_a} = \frac{(2\pi)^7}{2M} |\mathcal{T}_{fi}|^2 \, \delta^4(P_f - p_a) \, d\Phi \qquad (3.82)$$

and the decay width Γ follows by integration over $d\Phi$. For a decay process into n particles one has to insert the relation between $\mathcal{T}_{fi}$ and the matrix element $\mathcal{M}_{fi}$,

$$\mathcal{T}_{fi} = \left[\frac{1}{(2\pi)^{3/2}} \right]^{1+n} \mathcal{M}_{fi}.$$

For the important class of 2-particle decays the differential decay width becomes

$$d\Gamma = \frac{1}{8\pi^2 M} |\mathcal{M}_{fi}|^2 \, \delta^4(p_1 + p_2 - p_a) \frac{d^3 p_1}{2p_1^0} \frac{d^3 p_2}{2p_2^0}.$$

Writing $d^3 p_1$ in polar coordinates as specified in Eq. (3.74), and integrating over $d^3 p_2$ und $d|\vec{p}_1|$ one finds

$$d\Gamma = \frac{1}{8\pi^2 M} |\mathcal{M}_{fi}|^2 \frac{|\vec{p}|}{4M} \, d\Omega \qquad (3.83)$$

where the momentum $|\vec{p}_1| = |\vec{p}|$ is kinematically determined by

$$|\vec{p}| = \frac{1}{2M} \sqrt{\left(M^2 - m_1^2 - m_2^2\right)^2 - 4m_1^2 m_2^2}.$$

For particles of equal mass, $m_1 = m_2 \equiv m$, this expression simplifies to

$$|\vec{p}| = \frac{M}{2} \sqrt{1 - \frac{4m^2}{M^2}}.$$

The decay width Γ is obtained by integration over the angles,

$$\Gamma = \int d\Omega \, \frac{d\Gamma}{d\Omega}. \tag{3.84}$$

In case of particles with spin, the square $|\mathcal{M}_{fi}|^2$ has to be replaced by the spin-averaged square $\overline{|\mathcal{M}_{fi}|^2}$ in the expressions above, performing summation over the final-state helicities and averaging over the initial-state helicities.

3.7 On the Calculation of Cross Sections

For the practical calculations in squaring the matrix elements and summing over the helicities, several steps have to be done for which in particular the acquaintance with spinors and Dirac matrices is an indispensable prerequisite. In the following the most important rules are specified, which among others play an essential role in the application to processes like those shown in the examples of Sec. 3.5.

3.7.1 *Spinors in matrix elements*

For spinors $w = u$ or v and arbitrary 4×4 matrices Γ the following rules hold.

- Complex conjugates

$$(\overline{w}_1 \Gamma w_2)^* = \overline{w}_2 \overline{\Gamma} w_1 \quad \text{with } \overline{\Gamma} = \gamma^0 \Gamma^\dagger \gamma^0 \tag{3.85}$$

- Representation by traces

$$(\overline{w}_1 \Gamma_1 w_2)(\overline{w}_2 \Gamma_2 w_1) = \text{Tr}\left[(w_1 \overline{w}_1)\Gamma_1 (w_2 \overline{w}_2)\Gamma_2\right] \tag{3.86}$$

- Polarization sum

$$\sum_\sigma u_\sigma(p)\, \overline{u}_\sigma(p) = \not{p} + m$$

$$\sum_\sigma v_\sigma(p)\, \overline{v}_\sigma(p) = \not{p} - m \tag{3.87}$$

3.7.2 Traces over Dirac matrices

$$\mathrm{Tr}\big(\gamma^{\mu}\gamma^{\nu}\big) = 4\,g^{\mu\nu}$$

$$\mathrm{Tr}\big(\gamma^{\mu}\gamma^{\alpha}\gamma^{\nu}\gamma^{\beta}\big) = 4\left(g^{\mu\alpha}g^{\nu\beta} + g^{\mu\beta}g^{\nu\alpha} - g^{\mu\nu}g^{\alpha\beta}\right) \tag{3.88}$$

$$\mathrm{Tr}\big(\not{p}\gamma^{\mu}\not{q}\gamma^{\nu}\big) = 4\left(p^{\mu}q^{\nu} + p^{\nu}q^{\mu} - (pq)g^{\mu\nu}\right)$$

These rules follow by successive applications of the Dirac algebra (2.41). Furthermore, as another rule one has

$$\mathrm{Tr}\big(\text{odd number of Dirac matrices}\big) = 0.$$

3.7.3 Polarization sum for photons

The summation over the two physical transverse polarization states, according to the two helicity states of the photon, can be written as follows, with the help of $(r^{\mu}) = (k^{0}, -\vec{k}\,)$,

$$\sum_{\lambda=\pm 1} \epsilon_{\lambda}^{\mu*}(k)\,\epsilon_{\lambda}^{\nu}(k) = -g^{\mu\nu} + \frac{k^{\mu}r^{\nu} + k^{\nu}r^{\mu}}{(k\cdot r)}. \tag{3.89}$$

The second term, however, does not contribute to $\overline{|\mathcal{M}_{fi}|^{2}}$ in QED and can thus be omitted in practical calculations. The reason is gauge invariance whereby physical quantitites remain unchanged under a gauge transformation

$$A_{\mu}(x) \to A_{\mu}(x) + \partial_{\mu}\chi(x), \quad \chi(x) \quad \text{arbitrary,}$$

which in momentum space means $\epsilon_{\mu}(k) \to \epsilon_{\mu}(k) + k_{\mu}f(k)$. Any matrix element involving an external photon with momentum k is of the type

$$\mathcal{M} = \epsilon_{\mu}(k)\,T^{\mu}$$

and has to be invariant under the gauge transformation, which requires

$$k_{\mu}T^{\mu} = 0.$$

Hence, for the spin summation in $|\mathcal{M}|^{2}$ the second term in Eq. (3.89) yields zero.

Example 1. Cross section for $e^{+}e^{-} \to \mu^{+}\mu^{-}$

Starting from the matrix element (3.57) we demonstrate by means of the example of electron–positron annihilation into muon pairs how the calculation of the cross section is performed. The following abbreviating notation is used,

$$u = u_{\sigma_a}(p), \quad v = v_{\sigma_b}(q), \quad U' = U_{\sigma_1}(p'), \quad V' = V_{\sigma_2}(q').$$

Then the matrix element appears as follows,

$$\mathcal{M} = i\frac{e^2}{Q^2}\left(\bar{v}\gamma^\mu u\right)\left(\bar{U}'\gamma_\mu V'\right) \quad \text{with } Q^2 = (p+q)^2 = (p'+q')^2.$$

Squaring with the help of the rules (3.85) and (3.86) yields

$$|\mathcal{M}|^2 = \left(\frac{e^2}{Q^2}\right)^2 \left(\bar{v}\gamma^\mu u\right)\left(\bar{U}'\gamma_\mu V'\right)\cdot\left(\bar{v}\gamma^\nu u\right)^*\left(\bar{U}'\gamma_\nu V'\right)^*$$

$$= \left(\frac{e^2}{Q^2}\right)^2 \underbrace{\left(\bar{v}\gamma^\mu u\right)\left(\bar{u}\gamma^\nu v\right)}_{\mathrm{Tr}\left(v\bar{v}\,\gamma^\mu u\bar{u}\,\gamma^\nu\right)}\,\underbrace{\left(\bar{U}'\gamma_\mu V'\right)\left(\bar{V}'\gamma_\nu U'\right)}_{\mathrm{Tr}\left(U'\bar{U}'\gamma_\mu V'\bar{V}'\gamma_\nu\right)}.$$

The spin summation is performed with the help of the polarization sums (3.87),

$$\overline{|\mathcal{M}|^2} = \frac{1}{2}\cdot\frac{1}{2}\sum_{\sigma_a,\sigma_b}\sum_{\sigma_1,\sigma_2}|\mathcal{M}|^2$$

$$= \left(\frac{e^2}{Q^2}\right)^2\frac{1}{4}\mathrm{Tr}\left[(\slashed{q}-m_e)\gamma^\mu(\slashed{p}+m_e)\gamma^\nu\right]\mathrm{Tr}\left[(\slashed{p}'+m_\mu)\gamma_\mu(\slashed{q}'-m_\mu)\gamma_\nu\right].$$

The further steps are done in the high-energy approximation $|\vec{p}|,|\vec{q}| \gg m_\mu$ neglecting the masses m_e und m_μ. With the trace formulae (3.88) one obtains

$$\overline{|\mathcal{M}|^2} = \left(\frac{e^2}{Q^2}\right)^2 4\left[q^\mu p^\nu + q^\nu p^\mu - (pq)g^{\mu\nu}\right]\cdot\left[p'_\mu q'_\nu + p'_\nu q'_\mu - (p'q')g_{\mu\nu}\right].$$

By contraction of the Lorentz indices scalar products of 4-momenta occur which are usually expressed in terms of invariant kinematical variables (also denoted as Mandelstam variables). These are defined as follows,

$$\begin{aligned}s &= (p+q)^2 = (p'+q')^2 = \quad 2pq = \quad 2p'q',\\ t &= (p-p')^2 = (q-q')^2 = -2pp' = -2qq',\\ u &= (p-q')^2 = (q-p')^2 = -2pq' = -2qp'.\end{aligned} \qquad (3.90)$$

In terms of the Mandelstam variables the result of the spin summation is a simple expression,

$$\overline{|\mathcal{M}|^2} = 2e^4\frac{u^2+t^2}{s^2}.$$

Introducing the scattering angle θ in the CMS, the variables t and u can be rewritten as follows,

$$t = -\frac{s}{2}\left(1 - \cos\theta\right), \quad u = -\frac{s}{2}\left(1 + \cos\theta\right),$$

yielding

$$\overline{|\mathcal{M}|^2} = e^4\left(1 + \cos^2\theta\right).$$

The differential cross section in the CMS then follows directly from the expression given in Eq. (3.75),

$$\frac{d\sigma}{d\Omega} = \frac{1}{64\pi^2 s}\,\overline{|\mathcal{M}|^2} = \frac{\alpha^2}{4s}\left(1 + \cos^2\theta\right) \quad \text{with } \alpha = \frac{e^2}{4\pi}, \tag{3.91}$$

and the integrated cross section

$$\sigma = \int d\Omega\,\frac{d\sigma}{d\Omega} = \int_0^{2\pi} d\varphi \int_{-1}^{1} d\cos\theta\,\frac{d\sigma}{d\Omega}$$

gets the form

$$\boxed{\sigma = \frac{4\pi\alpha^2}{3s}}. \tag{3.92}$$

This cross section is a quantity of crucial importance for physics at e^+e^- colliders. In a quite similar way one obtains the cross section for the production of hadrons in electron–positron annihilation, which proceeds primarily via the production of quark–antiquark pairs,

$$\sigma(e^+e^- \to \text{hadrons}) = \sum_q \sigma(e^+e^- \to q\bar{q}). \tag{3.93}$$

According to the Feynman rules for general fermions, merely an extra charge factor Q_q has to be attached to the quark–photon vertex for each quark species, resulting in

$$\sigma(e^+e^- \to \text{hadrons}) = \frac{4\pi\alpha^2}{3s}\,N_C \sum_q Q_q^2 \tag{3.94}$$

(for energies $\sqrt{s} > 10\,\text{GeV}$ the sum extends over $q = u, d, s, c, b$ quarks). The colour factor $N_C = 3$ has been placed in addition since each quark q occurs in three different colour states. Historically the confirmation of this factor by measurements of the ratio

$$R = \frac{\sigma(e^+e^- \to \text{hadrons})}{\sigma(e^+e^- \to \mu^+\mu^-)} = N_C \sum_q Q_q^2 \tag{3.95}$$

was an essential step to prove the existence of the colour degree of freedom and thus towards establishing quantum chromodynamics.

Example 2.   Cross section for Compton scattering

The cross section for Compton scattering is customarily given in the laboratory frame, where the electron as the target is at rest. The matrix element can be found in Sec. 3.5. The notation for the momenta in the laboratory frame is as follows,

$$p_a = k = (\omega, \vec{k}), \quad p_1 = k' = (\omega', \vec{k}'), \quad p_b = (m_e, \vec{0}),$$

and the scattering angle is defined as the angle $\theta = \angle(\vec{k}, \vec{k}')$. The cross section follows according to Eq. (3.79),

$$\frac{d\sigma}{d\Omega} = \frac{1}{64\pi^2 m_e^2} \left(\frac{\omega'}{\omega}\right)^2 |\mathcal{M}|^2 \quad \text{with} \quad \frac{\omega'}{\omega} = \frac{m_e}{m_e + \omega(1 - \cos\theta)}.$$

Performing the spin summation yields the *Klein–Nishina formula*

$$\boxed{\frac{d\sigma}{d\Omega} = \frac{\alpha^2}{2m_e^2} \left(\frac{\omega'}{\omega}\right)^2 \left(\frac{\omega'}{\omega} + \frac{\omega}{\omega'} - \sin^2\theta\right)} \qquad (3.96)$$

for the differential Compton cross section. Note that for $\omega \ll m_e$ the classical cross section for Thomson scattering is recovered.

3.8   Precision Tests of QED

For the QED processes considered so far, the respective matrix elements were treated in lowest-order perturbation theory where the transition amplitudes for $2 \to 2$ scattering processes are of the order e^2 in the coupling constant. In general, perturbation theory provides an expansion in powers of the coupling constant e or the fine structure constant α, respectively,

$$\mathcal{M}_{fi} = \mathcal{M}^{(1)}(e^2) + \mathcal{M}^{(2)}(e^4) + \cdots = \alpha\,\mathcal{A}^{(1)} + \alpha^2\mathcal{A}^{(2)} + \cdots$$

Typical Feynman graphs of the next order, here for the example of $e^+ e^- \to \mu^+ \mu^-$,

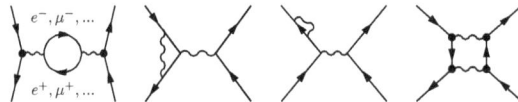

show the characteristic appearance of closed loops with a free momentum, i.e. not fixed by momentum conservation at the vertices. The list of Feynman rules given so far has to be extended accordingly for loop diagrams:

- Integration over the free momentum k within the closed loop,

$$\int \frac{d^4 k}{(2\pi)^4} \cdots$$

- Trace calculation for a closed fermion loop and an extra minus sign,

$$(-1)\operatorname{Tr}\{\cdots\}$$

Because of the small value of $\alpha \approx 1/137$ the orders $n > 1$ yield numerically small contributions compared to the leading order, rendering a kind of "corrections". Such corrections, however, are definitely relevant whenever the experimental accuracy is sufficiently high. Appropriate precision experiments that are sensitive to the higher-order terms in the predictions provide valuable tests of QED and verify the structure of QED as a quantum field theory.

3.8.1 *Precision tests at low energies*

At low energies two measureable effects are particularly prominent and historically had a significant impact to establish and consolidate QED as the correct theory of the electromagnetic interaction.

(i) **Lamb shift.** The Lamb shift [Lamb (1947)] is an effect in the spectrum of the hydrogen atom that is not contained in the prediction of the Dirac equation (see Sec. 2.4). The Lamb shift is defined as the upward shift of the $2s_{1/2}$ energy level with respect to the $2p_{1/2}$ level of the H-atom (Fig. 3.1), with the value $\Delta E_{\exp} = 4.37 \cdot 10^{-6}\,\text{eV} \sim 1058\,\text{MHz}$. Conventionally it is expressed in terms of the frequency of the transition. The measured shift could be explained as a quantum effect of QED [Bethe (1947); Karplus (1952)].

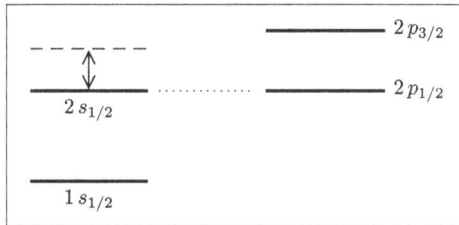

Fig. 3.1 Schematic energy levels of the H-atom including the Lamb shift (dashed line). The full lines correspond to the spectrum predicted by the Dirac equation.

The degeneracy with respect to the l quantum number, predicted by the Dirac equation for the energy levels in the Coulomb field, is removed by the Lamb shift. The reason is the modification of the Coulomb interaction between the electron and the proton by higher-order contributions of QED which are most pronounced for the s-levels: a deviation from the $1/r$ potential by the photon vacuum polarisation and a spatially extended charge distribution of the electron, given as the Fourier transformed of an electric form factor originating from the vertex correction, graphically visualized as follows:

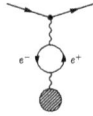

modification of the $1/r$ potential   form factor of the electron

(ii) **Anomalous magnetic moment.** The vertex correction provides also a magnetic form factor und thus a contribution to the magnetic dipole moment of the electron. Recall: the Dirac equation predicts for the electron (and the other leptons) a magnetic moment with the g-factor $g = 2$ (see Sec. 2.4).

By higher-order QED calculations additional terms contribute to the magnetic moment which lead to $g \neq 2$. The deviation from 2 is of the order 10^{-3}, the experiments, however, are of highest precision and allow to perform the most stringent tests of a physical theory at all. Customarily the deviation from the Dirac prediction for g,

$$a_e = \frac{g-2}{2} \tag{3.97}$$

is defined as the anomalous magnetic moment and used as the representative quantity for comparing theory and experiment. The contribution following from the vertex correction at one-loop order was calculated already 1948 with the result [Schwinger (1948)]

$$a_e = \frac{\alpha}{2\pi}$$

The currently most precise measured value is given by [Hanneke (2008)]

$$a_e = 0.001\,159\,652\,180\,73\,(28). \tag{3.98}$$

The theoretical prediction is based on the expansion

$$a_e = \frac{\alpha}{2\pi} + c_2 \left(\frac{\alpha}{\pi}\right)^2 + c_3 \left(\frac{\alpha}{\pi}\right)^3 + c_4 \left(\frac{\alpha}{\pi}\right)^4 + c_5 \left(\frac{\alpha}{\pi}\right)^5 + a_e^{\text{had}} + a_e^{\text{weak}}$$

calculated up to the 5-loop order in QED (including also μ and τ) [Aoyama (2019)]. The contributions from hadrons and from the weak interaction have to be taken into account as well [Jegerlehner (2019)],

$$a_e^{\text{had}} = 1.693(11) \cdot 10^{-12}$$

$$a_e^{\text{weak}} = 0.03053(23) \cdot 10^{-12}$$

as parts of the Standard Model. The prediction for a_e depends crucially on the precise value of α. With the most accurate value (from measurements with Cs atoms) [Parker (2018)]

$$\alpha^{-1} = 137.035\,999\,045\,(28)$$

the theory prediction based on the Standard Model is given by [Aoyama (2019)]

$$a_e = 0.001\,159\,652\,181\,61\,(23). \tag{3.99}$$

The agreement with the measured value (3.98) is remarkable; on the other hand, there is a deviation between theory and experiment of more than two standard deviations, which is currently an open issue.

The **anomalous magnetic moment of the muon**, a_μ, can be predicted in a similar way, though there are differences with respect to a_e originating from the bigger mass $m_\mu \approx 200\,m_e$, which in particular gives more weight to the non-QED effects, see [Jegerlehner (2017)] for a review. The QED contribution has been calculated up to the 5-loop order as in case of a_e, the hadronic and weak contributions amount to [Aoyama (2020)]

$$a_\mu^{\text{had}} = 69\,37\,(43) \cdot 10^{-11}$$

$$a_\mu^{\text{weak}} = 154\,(1) \cdot 10^{-11}$$

Altogether the Standard Model prediction is given by [Aoyama (2020)]

$$a_\mu = 0.001\,165\,918\,10\,(43). \tag{3.100}$$

The most precise experimental value until now, measured at the Brookhaven National Laboratory [Bennett (2006)],

$$a_\mu = 0.001\,165\,920\,91\,(63) \tag{3.101}$$

is 3.7 standard deviations above the theory prediction. The difference is currently not understood and has motivated lots of spec-
ulations about further contributions that could be associated with
empirically unknown new physics beyond the Standard Model. A new
experiment at the Fermi National Laboratory [Grange (2015)] for measuring a_μ with a four times higher accuracy will help to clarify the situation. A first result [Abi (2021)] is compatible with the value (3.101),
$a_\mu^{\mathrm{FNAL}} = 0.001\ 165\ 920\ 40\ (54)$.

3.9 Addendum: Coulomb Gauge and Feynman Graphs

The free radiation field is described by the Hamiltonian (3.13), expressed in
terms of the vector potential $\vec{A}$ in Coulomb gauge. In case of an interaction
with a current j^μ there is, in addition to the $\vec{j}\cdot\vec{A}$ term, also the Coulomb
interaction $j^0 A_0$ with the scalar potential $A^0 = \phi$. In Eq. (3.43) it was proposed to choose the interaction part of the Hamiltonian as $j^\mu A_\mu$ including
the Coulomb interaction, and the proper free field part H_0^{em} as containing only the transverse radiation field. A closer inspection shows that this
handling is formally not quite correct since the restriction to the physical
degrees of freedom using the Coulomb gauge is not covariant. A more rig-
orous treatment, however, leads at the end to the same Feynman rules and
Feynman graphs, so that the seemingly covariant approach of Sec. 3.4 can
be regarded as a simplified way towards the same result. This section provides additional information and outlines the formal steps in the procedure.

The QED Hamiltonian for the electromagnetic field and the interaction
reads as follows (see Sec. 4.3),

$$H = \frac{1}{2} \int d^3x \left(\vec{E}^2 + \vec{B}^2\right) - e \int d^3x\, \vec{j}\cdot\vec{A} \qquad (3.102)$$

where $\vec{E} = \vec{E}_{\mathrm{T}} + \vec{E}_{\mathrm{L}}$ in Coulomb gauge, with

$$\vec{E}_{\mathrm{T}} = -\frac{\partial \vec{A}}{\partial t}, \quad \vec{E}_{\mathrm{L}} = -\nabla\phi,$$

contains the transverse radiation field $\vec{E}_{\mathrm{T}}$ and the longitudinal Coulomb
field $\vec{E}_{\mathrm{L}}$, so that[2]

$$\int d^3x\, \vec{E}^2 = \int d^3x\, \vec{E}_{\mathrm{T}}^2 + \int d^3x\, \vec{E}_{\mathrm{L}}^2.$$

---

[2]Integrals over a divergence ocurring in some intermediate steps are omitted as irrelevant surface terms.

The integral over the $\vec{E}_\mathrm{L}$ part represents the energy of the Coulomb interaction. Because of

$$\nabla \cdot \vec{E}_\mathrm{L} = -\Delta\phi = \rho = e\, j^0$$

and with $\phi(x)$ written as solution of the Poisson equation, it reads

$$\frac{1}{2} \int d^3x\, \vec{E}_\mathrm{L}^2 = \frac{1}{2} \int d^3x \left(\nabla\phi\right)^2 = -\frac{1}{2} \int d^3x \left(\phi\,\Delta\phi\right)$$

$$= \frac{e^2}{8\pi} \int d^3x\, j^0(\vec{x},t) \int d^3y\, \frac{j^0(\vec{y},t)}{|\vec{x}-\vec{y}|}.$$

The Hamiltonian (3.102) thus becomes

$$H = H_0^{\mathrm{em}} + H_{\mathrm{int}} = H_0^{\mathrm{em}} + H_{\mathrm{int}}^{\mathrm{C}} + H_{\mathrm{int}}^{\mathrm{trans}} \tag{3.103}$$

where

$$H_0^{\mathrm{em}} = \frac{1}{2} \int d^3x \left(\vec{E}_\mathrm{T}^2 + \vec{B}^{\,2}\right) \tag{3.104}$$

corresponds to the free radiation field; the interaction terms are given by the transverse contribution

$$H_{\mathrm{int}}^{\mathrm{trans}} = -e \int d^3x\, \vec{j} \cdot \vec{A} \tag{3.105}$$

and by the instantaneous Coulomb interaction

$$H_{\mathrm{int}}^{\mathrm{C}} = \frac{e^2}{8\pi} \int d^3x \int d^3y\, \frac{j^0(\vec{x},t)\, j^0(\vec{y},t)}{|\vec{x}-\vec{y}|}. \tag{3.106}$$

In matrix elements, the contribution from $H_{\mathrm{int}}^{\mathrm{C}}$ compensates the instantaneous part in the transverse photon propagator (3.31) from internal photon lines, with the consequence that H_{int} in Eq. (3.43) with the full photon propagator (3.28) effectively yields the same result. The polarization sum $g_{\mu\nu}$ means that formally the summation extends over four polarization vectors; two of them are unphysical, they correspond to a longitudinal and a timelike polarization. The timelike polarization refers to the Coulomb interaction, the longitudinal one does not contribute to the matrix elements because of current conservation or gauge invariance, respectively.

To make this equivalence explicit we first consider the Coulomb interaction part in the S-operator (3.51),

$$
\begin{aligned}
-i \int_{-\infty}^{\infty} dt\, H_{\text{int}}^{C} &= -i \int_{-\infty}^{\infty} dt\, \frac{e^2}{8\pi} \int d^3x \int d^3y\, \frac{j^0(\vec{x},t)\, j^0(\vec{y},t)}{|\vec{x}-\vec{y}|} \\
&= -i \frac{e^2}{8\pi} \int d^4x \int d^4y\, \delta(x^0-y^0)\, \frac{j^0(x)\, j^0(y)}{|\vec{x}-\vec{y}|} \\
&= -i \frac{e^2}{16\pi^2} \int d^4x \int d^4y \int dQ^0\, e^{-iQ^0(x^0-y^0)}\, \frac{j^0(x)\, j^0(y)}{|\vec{x}-\vec{y}|}.
\end{aligned}
$$

After insertion of

$$
\frac{1}{|\vec{x}-\vec{y}|} = \int \frac{d^3Q}{2\pi^2}\, \frac{e^{i\vec{Q}(\vec{x}-\vec{y})}}{\vec{Q}^2}
$$

one obtains

$$
\begin{aligned}
-i \int_{-\infty}^{\infty} dt\, H_{\text{int}}^{C} &= -i \frac{1}{2} \int \frac{d^4Q}{(2\pi)^4} \int d^4x \int d^4y\, e^{-iQ(x-y)}\, \frac{j^0(x)\, j^0(y)}{\vec{Q}^2} \\
&= -i \frac{1}{2} \int \frac{d^4Q}{(2\pi)^4} \int d^4x \int d^4y\, e^{-iQ(x-y)}\, \frac{j^\mu(x)\, \eta_\mu \eta_\nu\, j^\nu(y)}{\vec{Q}^2}
\end{aligned}
\tag{3.107}
$$

with $(\eta_\mu) = (1,0,0,0)$.

The further procedure is demonstrated by means of the example
$e^+e^- \to \mu^+\mu^-$, following Sec. 3.5. With the current formed by the $e^\pm$
and $\mu^\pm$ fields ψ and Ψ,

$$
j^\mu = \overline{\psi}\gamma^\mu\psi + \overline{\Psi}\gamma^\mu\Psi,
\tag{3.108}
$$

the numerator of (3.107) contains the sum

$$
\left(\overline{\psi}\gamma^\mu\psi\right)\left(\overline{\Psi}\gamma^\nu\Psi\right) + \left(\overline{\Psi}\gamma^\mu\Psi\right)\left(\overline{\psi}\gamma^\nu\psi\right)
$$

where each term provides the same contribution to the matrix element, thus
compensating the factor $1/2$. Inserting the Fourier expansions of the field
operators and computing the matrix element of Eq. (3.107) between the
states $\langle f|$ and $|i\rangle$ yields the Coulomb contribution to $\mathcal{M}_{fi}$, with $Q = p+q$,

$$
\mathcal{M}_{fi}^{C} = \left(\overline{v}\, ie\gamma^\mu u\right) \left(\frac{i\eta_\mu\eta_\nu}{\vec{Q}^2}\right) \left(\overline{U}\, ie\gamma^\nu V\right).
$$

The other contribution arising from the transverse part of H_{int} in Eq. (3.103) contains the transverse photon propagator with the polarization sum (3.31),

$$i\,D_{\mu\nu}^{\text{trans}}(Q) = \frac{i}{Q^2 + i\varepsilon}\left(-g_{\mu\nu} - \frac{Q_\mu Q_\nu - (Q_\mu \eta_\nu + Q_\nu \eta_\mu)(\eta Q) + \eta_\mu \eta_\nu \, Q^2}{\vec{Q}^2}\right)$$

(3.109)

and is given by[3]

$$\mathcal{M}_{fi}^{\text{trans}} = (\overline{v}\,ie\gamma^\mu u)\,iD_{\mu\nu}^{\text{trans}}(Q)\,(\overline{U}ie\gamma^\nu V)$$

$$= (\overline{v}\,ie\gamma^\mu u)\left(-\frac{ig_{\mu\nu}}{Q^2} - \frac{i\eta_\mu \eta_\nu}{\vec{Q}^2}\right)(\overline{U}ie\gamma^\nu V).$$

(3.110)

In the sum the Coulomb contribution compensates the $\eta_\mu \eta_\nu$ term of the propagator and one obtains

$$\mathcal{M}_{fi} = \mathcal{M}_{fi}^{\text{C}} + \mathcal{M}_{fi}^{\text{trans}} = (\overline{v}\,ie\gamma^\mu u)\left(\frac{-ig_{\mu\nu}}{Q^2}\right)(\overline{U}ie\gamma^\nu V)$$

which is the same expression as in Eq. (3.57).

3.9.1 Classical solution for the radiation field

The transverse propagator in Eq. (3.109) has been obtained by evaluating the two-point function (3.29) of the quantized photon field with exclusively transverse polarization vectors. It remains to be shown that this expression is also the Green function of the classical wave equation for the transverse radiation field in Coulomb gauge.

The classical Maxwell equations for the potentials ϕ and $\vec{A}$ in Coulomb gauge follow from Eq. (1.84),

$$\Delta\phi = -e\,j^0,$$

(3.111)

$$\Box\vec{A} = e\vec{j} - \nabla\frac{\partial\phi}{\partial t} \equiv e\vec{j}_{\text{T}}.$$

(3.112)

The solution of the first equation (3.111) is given by the instantaneous Coulomb potential,

$$\phi(\vec{x}, t) = \frac{e}{4\pi}\int d^3y\,\frac{j^0(\vec{y}, t)}{|\vec{x} - \vec{y}|}$$

(3.113)

---

[3]Since $D_{0\nu}^{\text{trans}} = 0$ only the space components of the current contribute, as required by (3.105).

and determines the inhomegeneity $\vec{j}_T$ (the transverse current) in the wave equation (3.112) for the transverse radiation field. The inhomogeneous wave equation can be solved by means of the Green function (3.26), yielding

$$A_k(x) = e \int d^4y \, D_{kl}(x-y) \, j_T^l(y) \tag{3.114}$$

which can be rewritten in terms of the transverse Green function as follows,

$$A_k(x) = e \int d^4y \, D_{kl}^{\text{trans}}(x-y) \, j^l(y), \tag{3.115}$$

with

$$D_{kl}^{\text{trans}}(x-y) = \int \frac{d^4Q}{(2\pi)^4} \, e^{-iQ(x-y)} \, D_{kl}^{\text{trans}}(Q) \tag{3.116}$$

and

$$D_{kl}^{\text{trans}}(Q) = \frac{1}{Q^2 + i\varepsilon} \left(-g_{kl} - \frac{Q_k Q_l}{\vec{Q}^2} \right). \tag{3.117}$$

It obeys the transversality condition

$$Q^k D_{kl}^{\text{trans}}(Q) = 0$$

and thus assures that the radiation field $\vec{A}$ in Eq. (3.115) fulfills the Coulomb gauge condition $\nabla \cdot \vec{A} = 0$. The physical meaning of D_{kl}^{trans} is that it propagates only the transverse polarizations of the electromagnetic field arising from a given source.

The transverse Green function (3.117) is recognized as the 3-dimensional space part of the transverse photon propagator in Eq. (3.109), with $D_{0\nu}^{\text{trans}} = 0$. In other words, $D_{\mu\nu}^{\text{trans}}$ is the 4-dimensional notation for the transverse Green function of the wave equation.

Chapter 4

Lagrangians and Symmetries

The Lagrangian formulation of mechanics and field theory is a particularly suitable framework for the description of physical systems. A multitude of reasons can be given for the Lagrangian approach.

- The Lagrangian L is the fundamental dynamical quantity of a physical system.
- The action $S = \int dt L$ is a Lorentz-invariant quantity, with the consequence that the equations of motion are covariant.
- The equations of motion follow via Hamilton's principle that the variation of the action is zero, $\delta S = 0$. This principle is universal and can be applied to all areas of physics.
- Symmetries of a physical system can be formulated in an easy way as symmetries of L or S, respectively, with direct consequences like the existence of conservation laws (Noether theorem). This concerns space–time symmetries as well as internal symmetries.
- Modern theories of the fundamental interactions are based on symmetries and can be constructed within the Lagrangian formalism according to a set of simple rules.

The following sections on topics of Langrangian field theory apply to classical fields and can also be transferred to quantum fields.

4.1 Lagrangian Formalism for Fields

We start by recalling the Lagrangian method in classical mechanics. A system of pointlike particles with coordinates q_k and velocities $\dot{q}_k$ is described by a Lagrangian $L = L(q_1, \ldots, \dot{q}_1, \ldots)$. The variation of the action

$$\delta S = 0, \qquad S = \int_{t_1}^{t_2} dt\, L,$$

gives rise to the equations of motion in terms of the Euler-Lagrange equations,

$$\frac{\partial L}{\partial q_k} - \frac{d}{dt}\left(\frac{\partial L}{\partial \dot{q}_k}\right) = 0.$$

The transition to field theory is done by replacing the discrete index k by the continuous index $\vec{x}$, the space coordinates of a field ϕ, according to

$$q_k(t) \to \phi(t, \vec{x}) \equiv \phi(x)$$

$$\dot{q}_k(t) \to \partial_\mu \phi(x)$$

$$L(q_k, \dot{q}_k) \to L = \int d^3x\, \mathcal{L}\big(\phi(x), \partial_\mu \phi(x)\big).$$

The Lagrangian appears in field theory as an integral over a *Lagrangian density* $\mathcal{L}(\phi, \partial_\mu \phi)$ (conventionally also denoted as *Lagrangian*), which is invariant under Lorentz transformations. The action becomes

$$S = \int_{-\infty}^{\infty} dt \int d^3x\, \mathcal{L} = \int d^4x\, \mathcal{L} \tag{4.1}$$

and is invariant as well because the volume element d^4x is invariant. A variation of the field,

$$\phi \to \phi + \delta\phi, \qquad \text{with} \quad \phi(x) \to 0 \quad \text{for} \quad x^\mu \to \pm\infty,$$

$$\partial_\mu \phi \to \partial_\mu \phi + \delta(\partial_\mu \phi) = \partial_\mu \phi + \partial_\mu(\delta\phi)$$

generates a variation of the action, which is set to zero according to Hamilton's principle,

$$\delta S = \int d^4x \left[\frac{\partial \mathcal{L}}{\partial \phi} + \frac{\partial \mathcal{L}}{\partial(\partial_\mu \phi)}\, \partial_\mu(\delta\phi)\right]$$

$$= \int d^4x\, \delta\phi \left[\frac{\partial \mathcal{L}}{\partial \phi} - \partial_\mu \left(\frac{\partial \mathcal{L}}{\partial(\partial_\mu \phi)}\right)\right] = 0.$$

Since $\delta\phi(x)$ is arbitrary, this implies the equation of motion as a field equation for ϕ

$$\boxed{\frac{\partial \mathcal{L}}{\partial \phi} - \partial_\mu \left(\frac{\partial \mathcal{L}}{\partial(\partial_\mu \phi)}\right) = 0} . \tag{4.2}$$

In case of several fields ϕ_j one obtains for each field an equation of this kind. Therein, $\phi_j(x)$ is a generic notation comprising different types of fields, i.e. ϕ_j can be: scalar fields ϕ and $\phi^\dagger$, the components A_ν of vector fields, or the components ψ_a and $\overline{\psi}_a$ of spinor fields.

Specific examples of Lagrangians and equations of motion for free fields shall illustrate the technique and simultaneously provide the basis for applications to the theory of fundamental interactions.

4.1.1 *Scalar field*

A neutral scalar field $\phi(x)$, with $\phi = \phi^\dagger$, is described by the Lagrangian (density)

$$\mathcal{L} = \frac{1}{2}\left(\partial_\nu \phi\right)\left(\partial^\nu \phi\right) - \frac{m^2}{2}\phi^2. \tag{4.3}$$

With the derivatives

$$\frac{\partial \mathcal{L}}{\partial(\partial_\mu \phi)} = \partial^\mu \phi, \qquad \frac{\partial \mathcal{L}}{\partial \phi} = -m^2 \phi$$

one obtains the equation of motion for ϕ, according to Eq. (4.2), as follows

$$-m^2 \phi - \partial_\mu\left(\partial^\mu \phi\right) = 0 \qquad \text{or} \qquad \left(\Box + m^2\right)\phi = 0, \tag{4.4}$$

which is the Klein–Gordon equation.

A complex scalar field $\phi \neq \phi^\dagger$ corresponds to two real fields ϕ_1, ϕ_2 according to

$$\phi = \frac{1}{\sqrt{2}}\left(\phi_1 + i\phi_2\right), \qquad \phi^\dagger = \frac{1}{\sqrt{2}}\left(\phi_1 - i\phi_2\right),$$

so that

$$\mathcal{L} = \mathcal{L}(\phi_1, \partial_\nu \phi_1) + \mathcal{L}(\phi_2, \partial_\nu \phi_2) = \left(\partial_\nu \phi^\dagger\right)\left(\partial^\nu \phi\right) - m^2 \phi^\dagger \phi. \tag{4.5}$$

With the derivatives with respect to $\phi^\dagger$,

$$\frac{\partial \mathcal{L}}{\partial(\partial_\mu \phi^\dagger)} = \partial^\mu \phi, \qquad \frac{\partial \mathcal{L}}{\partial \phi^\dagger} = -m^2 \phi,$$

one obtains the equation of motion for ϕ as a Klein–Gordon equation like (4.4). Doing the derivatives with respect to ϕ yields a Klein–Gordon-Gleichung for the adjoint field $\phi^\dagger$.

4.1.2 *Dirac field*

The Lagrangian for a Dirac field ψ is given by

$$\mathcal{L} = \overline{\psi}\left(i\gamma^{\nu}\partial_{\nu} - m\right)\psi, \tag{4.6}$$

or expressed in terms of the components ψ_a and $\overline{\psi}_a$ of the spinor ψ and of the adjoint spinor $\overline{\psi}$,

$$\mathcal{L} = \sum_{a,b} \overline{\psi}_a\, i\gamma^{\nu}_{ab}\, \partial_{\nu}\psi_b - m\sum_a \overline{\psi}_a\psi_a. \tag{4.7}$$

With the derivatives

$$\frac{\partial\mathcal{L}}{\partial(\partial_{\mu}\overline{\psi}_c)} = 0, \qquad \frac{\partial\mathcal{L}}{\partial(\overline{\psi}_c)} = \sum_{a,b}\delta_{ac}\,i\gamma^{\nu}_{ab}\,\partial_{\nu}\psi_b - m\sum_a \delta_{ac}\psi_a,$$

one obtains from the general expression (4.2), by specification of the generic field ϕ according to $\phi \to \overline{\psi}_c$, the equations of motion for the spinor components ψ_c in the following way,

$$\frac{\partial\mathcal{L}}{\partial\overline{\psi}_c} = 0 = \sum_b i\gamma^{\nu}_{cb}\,\partial_{\nu}\psi_b - m\psi_c. \tag{4.8}$$

This is the Dirac equation for ψ written in components.

In the usual matrix notation Eq. (4.8) appears as the familiar Dirac equation,

$$\left(i\gamma^{\nu}\partial_{\nu} - m\right)\psi = 0.$$

Taking the derivatives with respect to ψ_c yields the adjoint Dirac equation.

4.1.3 *Vector field*

Massless field. The Lagrangian for a massless vector field like the electromagnetic field A_{ν} with the field strength tensor $F_{\mu\nu} = \partial_{\mu}A_{\nu} - \partial_{\nu}A_{\mu}$ is given by

$$\mathcal{L} = -\frac{1}{4}F_{\rho\sigma}F^{\rho\sigma}. \tag{4.9}$$

Expressed in terms of the field strengths $\vec{E}$ and $\vec{B}$, $\mathcal{L}$ reads as follows,

$$\mathcal{L} = \frac{1}{2}\left(\vec{E}^{\,2} - \vec{B}^{\,2}\right) \tag{4.10}$$

which is not manifestly covariant. The equations of motion for A_ν,

$$\partial_\mu \left(\frac{\partial \mathcal{L}}{\partial(\partial_\mu A_\nu)} \right) = 0,$$

with

$$\frac{\partial \mathcal{L}}{\partial(\partial_\mu A_\nu)} = -\partial^\mu A^\nu + \partial^\nu A^\mu = -F^{\mu\nu}$$

get the explicit form

$$\partial_\mu F^{\mu\nu} = 0 \tag{4.11}$$

corresponding to Maxwell's equations for the free electromagnetic field.

Massive field. Of special interest is also the massive vector field, describing particles with spin 1 and mass m. The respective Lagrangian is obtained by adding a mass term to Eq. (4.9),

$$\mathcal{L} = -\frac{1}{4} F_{\rho\sigma} F^{\rho\sigma} + \frac{m^2}{2} A_\rho A^\rho, \tag{4.12}$$

and the field equation becomes the *Proca equation*

$$\partial_\mu F^{\mu\nu} + m^2 A^\nu = 0 = \left[(\Box + m^2) g^\nu_\mu - \partial_\mu \partial^\nu \right] A^\mu. \tag{4.13}$$

By differentiation follows the transversality condition $\partial_\mu A^\mu = 0$,

$$\partial_\nu \left[(\Box + m^2) g^\nu_\mu - \partial_\mu \partial^\nu \right] A^\mu = m^2 \partial_\mu A^\mu = 0, \tag{4.14}$$

which in contrast to the electromagnetic field is not an option but a necessary consequence. Hence, equivalent to Eq. (4.13) is the field equation

$$\left(\Box + m^2 \right) A^\nu = 0 \tag{4.15}$$

for each component of the vector field with $\partial_\nu A^\nu = 0$ as a side condition restricting the number of independent degrees of freedom of A^ν to three. The independent degrees of freedom correspond to the three polarizations, or equivalently to the three helicities $\lambda = \pm 1, 0$ of the massive spin-1 particles.

4.2 Space–Time Symmetries

Symmetries of a physical system under specific transformations appear within the Lagrangian formalism as an invariance of the action, which essentially is also an invariance of $\mathcal{L}$. Symmetry under space–time transformations means invariance of S or $\mathcal{L}$, respectively, under Poincaré transformations (Lorentz transformations plus translations in space and time).

4.2.1 *Translational invariance and 4-momentum*

Invariance under space–time translations imply conservation laws for energy and momentum. The crucial quantity therefore is the energy-momentum tensor

$$T^{\mu\nu} = \sum_j \frac{\partial \mathcal{L}}{\partial(\partial_\mu \phi_j)}(\partial^\nu \phi_j) - g^{\mu\nu}\,\mathcal{L} \qquad (4.16)$$

with the property

$$\partial_\mu T^{\mu\nu} = 0, \qquad (4.17)$$

which summarizes energy and momentum conservation. The *energy density* of the field is given by T^{00}, the *momentum density* by T^{0k}. They determine the mechanical observables of the field as follows,

- field energy, respectively Hamiltonian: $H = \int d^3x\, T^{00}$
- field momentum: $P^k = \int d^3x\, T^{0k}$

On the derivation of the energy-momentum tensor. For simplicity only a single field $\phi(x)$ is considered. The Lagrangian as a scalar quantity (scalar field) is invariant under Poincaré transformations $x^\mu \to x'^\mu$. The Lagrangian depends on x exclusively via the fields and their derivatives, as a composed function

$$\mathcal{L}(x) = \mathcal{L}\big(\phi(x), \partial_\mu \phi(x)\big). \qquad (4.18)$$

The invariance of $\mathcal{L}$ means

$$\mathcal{L}'(x') = \mathcal{L}(x) = \mathcal{L}(x(x')), \qquad (4.19)$$

i.e. the transformation occurs by substitution of the variables in $\mathcal{L}$. Consider now an infinitesimal translation

$$x'^\nu = x^\nu - \epsilon^\nu \quad \text{or inverse} \quad x^\nu = x'^\nu + \epsilon^\nu. \qquad (4.20)$$

By substitution of the variables according to (4.19) one obtains

$$\mathcal{L}(x(x')) = \mathcal{L}(x' + \epsilon). \qquad (4.21)$$

To simplify the notation we write x instead of x' in the following. Expansion to first order yields

$$\mathcal{L}(x + \epsilon) - \mathcal{L}(x) = \epsilon^\nu\, \partial_\nu \mathcal{L}(x). \qquad (4.22)$$

On the other hand, the expansion for $\mathscr{L}$ as a composed function is given by

$$
\begin{aligned}
\mathscr{L}(x+\epsilon) &- \mathscr{L}(x) \\
&= \epsilon^\nu \left[\frac{\partial \mathcal{L}}{\partial \phi}(\partial_\nu \phi) + \left(\frac{\partial \mathcal{L}}{\partial(\partial_\mu \phi)} \right) \partial_\nu (\partial_\mu \phi) \right]_{\phi=\phi(x)} \\
&= \epsilon^\nu \left[\frac{\partial \mathcal{L}}{\partial \phi}(\partial_\nu \phi) + \left(\frac{\partial \mathcal{L}}{\partial(\partial_\mu \phi)} \right) \partial_\mu (\partial_\nu \phi) \right] \\
&= \epsilon^\nu \left[\frac{\partial \mathcal{L}}{\partial \phi}(\partial_\nu \phi) - \partial_\mu \left(\frac{\partial \mathcal{L}}{\partial(\partial_\mu \phi)} \right) (\partial_\nu \phi) + \partial_\mu \left(\frac{\partial \mathcal{L}}{\partial(\partial_\mu \phi)}(\partial_\nu \phi) \right) \right]
\end{aligned}
\tag{4.23}
$$

(it is understood that the insertion indicated in the first line is always done). Because of the equation of motion the first two terms add up to zero and the result is

$$
\mathscr{L}(x+\epsilon) - \mathscr{L}(x) = \epsilon^\nu \, \partial_\mu \left(\frac{\partial \mathcal{L}}{\partial(\partial_\mu \phi)}(\partial_\nu \phi) \right).
\tag{4.24}
$$

In combination with Eq. (4.22) one obtains

$$
\epsilon^\nu \, \partial_\mu \left(\frac{\partial \mathcal{L}}{\partial(\partial_\mu \phi)}(\partial_\nu \phi) - g^\mu_\nu \, \mathcal{L} \right) = 0.
\tag{4.25}
$$

In the last step the notation for $\mathscr{L}$ was changed back to the explicitly composed form (4.18). Since ϵ^ν are arbitrary and independent, Eq. (4.25) implies for each ν the condition

$$
\partial_\mu \left(\frac{\partial \mathcal{L}}{\partial(\partial_\mu \phi)}(\partial_\nu \phi) - g^\mu_\nu \, \mathcal{L} \right) = \partial_\mu T^\mu_{\ \nu} = 0,
\tag{4.26}
$$

which is the conservation law for the energy-momentum tensor

$$
T^\mu_{\ \nu} = \frac{\partial \mathcal{L}}{\partial(\partial_\mu \phi)}(\partial_\nu \phi) - g^\mu_{\ \nu} \, \mathcal{L}.
\tag{4.27}
$$

In case of several fields or field components ϕ_j, the expansion (4.23) has to be extended by summing over j, and the expression (4.16) for $T^{\mu\nu}$ is recovered.

In the following, specific examples are presented for the various types of fields encountered so far.

4.2.1.1 *Scalar field*

For a neutral scalar field ϕ the Lagrangian given in Eq. (4.3) with the derivative

$$\frac{\partial \mathcal{L}}{\partial(\partial_0 \phi)} = \partial^0 \phi = \partial_0 \phi$$

leads to the Hamiltonian density

$$T^{00} = (\partial_0 \phi)^2 - \mathcal{L} \tag{4.28}$$

and the momentum density

$$T^{0k} = -T^0{}_k = -(\partial^0 \phi)(\partial_k \phi) \tag{4.29}$$

and thus to the Hamiltonian

$$H = \int d^3 x\, T^{00} = \frac{1}{2} \int d^3 x\, \left[(\partial_0 \phi)^2 + |\nabla \phi|^2 + m^2 \phi^2\right] \tag{4.30}$$

and to the momentum of the field,

$$P^k = \int d^3 x\, T^{0k}, \qquad \vec{P} = \int d^3 x\, \left(-i\partial_0 \phi\right)\left(-i\nabla \phi\right). \tag{4.31}$$

For a quantum field, the Fourier expansion of ϕ inserted yields the representation of H and $\vec{P}$ in terms of the number operators as given in Sec. 2.1.

4.2.1.2 *Dirac field*

The generic fields ϕ_j in Eq. (4.16) are, in case of a Dirac field, the spinor components $\psi_a, \overline{\psi}_a$ of the Dirac spinor ψ and its adjoint $\overline{\psi}$. The corresponding Dirac Lagrangian has been specified in Eq. (4.6). With the derivatives

$$\frac{\partial \mathcal{L}}{\partial(\partial_\mu \overline{\psi}_c)} = 0, \qquad \frac{\partial \mathcal{L}}{\partial(\partial_\mu \psi_c)} = \sum_a \overline{\psi}_a i\gamma^\mu_{ac} \tag{4.32}$$

the energy density follows according to (4.16),

$$T^{00} = T^0_0 = \sum_c \frac{\partial \mathcal{L}}{\partial(\partial_0 \psi_c)} \partial_0 \psi_c - g^0_0 \mathcal{L} = \sum_{a,c} \overline{\psi}_a i\gamma^0_{ac} \partial_0 \psi_c - \mathcal{L} \tag{4.33}$$

$$= \overline{\psi}\, i\gamma^0 \partial_0 \psi - \overline{\psi}\left[i\gamma^0 \partial_0 \psi + i\gamma^k \partial_k \psi - m\psi\right]$$

$$= \overline{\psi}\left[-i\gamma^k \partial_k \psi + m\psi\right]$$

$$= \psi^\dagger \gamma^0 \left[-i\gamma^k \partial_k \psi + m\psi\right]$$

$$= \psi^\dagger \left[-i\alpha^k \partial_k \psi + \beta m\psi\right]$$

$$= \psi^\dagger \left[-i\vec{\alpha} \cdot \nabla + \beta m\right]\psi,$$

making use of the notation from Sec. 2.2,

$$\gamma^0 = \beta, \quad \gamma^k = \gamma^0 \alpha^k, \quad \vec{\alpha} = (\alpha^1, \alpha^2, \alpha^3).$$

Integrating T^{00} over d^3x yields the Hamiltonian of the Dirac field as specified in Sec. 2.2.

By analogous steps one obtains the momentum density T^{0k} of the field,

$$T^{0k} = -T^0{}_k = -\sum_c \frac{\partial \mathcal{L}}{\partial(\partial_0 \psi_c)} \partial_k \psi_c = -\sum_{a,c} \overline{\psi}_a i \gamma^0_{ac} \partial_k \psi_c$$

$$= -\overline{\psi} i \gamma^0 \partial_k \psi = \psi^\dagger \big(-i \partial_k \big) \psi,$$

and thus the momentum operator is given by

$$P^k = \int d^3x \, T^{0k}, \qquad \vec{P} = \int d^3x \, \psi^\dagger \big(-i\nabla \big) \psi \equiv \int d\vec{P} \qquad (4.34)$$

as already mentioned in Sec. 2.2.

4.2.1.3 *Electromagnetic radiation field*

The Lagrangian of the electromagnetic field has been specified in Eqs. (4.9) and (4.10). For a free radiation field $(A^\mu) = (A^0, \vec{A})$ in radiation gauge,

$$A^0 = 0, \quad \nabla \cdot \vec{A} = 0,$$

the field strengths are given by

$$\vec{E} = -\partial_0 \vec{A}, \quad \vec{B} = \nabla \times \vec{A}, \qquad (4.35)$$

and the corresponding Lagrangian can be written as follows,

$$\mathcal{L} = \frac{1}{2} \left[(\partial_0 \vec{A})^2 - (\nabla \times \vec{A})^2 \right]. \qquad (4.36)$$

Here, the generic fields ϕ_j in Eq. (4.16) correspond to the transverse components A^k of the vector potential. With the derivatives

$$\frac{\partial \mathcal{L}}{\partial(\partial_0 A^k)} = \partial_0 A^k$$

the energy density of the field

$$T^{00} = T^0{}_0 = \sum_k \frac{\partial \mathcal{L}}{\partial(\partial_0 A^k)} (\partial_0 A^k) - \mathcal{L} = \sum_k (\partial_0 A^k)^2 - \mathcal{L} \qquad (4.37)$$

$$= \frac{1}{2} \left[(\partial_0 \vec{A})^2 + (\nabla \times \vec{A})^2 \right] = \frac{1}{2} \left[\vec{E}^2 + \vec{B}^2 \right]$$

arises in the form known from classical electrodynamics.

The momentum density

$$T^{0k} = -T^0{}_k = -\sum_l \frac{\partial \mathcal{L}}{\partial(\partial_0 A^l)} \partial_k A^l \tag{4.38}$$

$$= -\sum_l (\partial_0 A^l)(\partial_k A^l) = -(\partial^0 A_l)(\partial^k A^l),$$

however, is different from the Poynting vector

$$\vec{S} = \vec{E} \times \vec{B} \tag{4.39}$$

which is known from electrodynamics to represent the density of the field momentum. The difference can easily be seen by means of Eq. (4.35),

$$(\vec{E} \times \vec{B})_k = -\epsilon_{klm}(\partial_0 A^l)\,\epsilon_{mjn}(\partial_j A^n) \tag{4.40}$$

$$= -(\delta_{kj}\delta_{ln} - \delta_{kn}\delta_{lj})\,(\partial_0 A^l)(\partial_j A^n)$$

$$= -(\partial_0 A^l)(\partial_k A^l) + (\partial_0 A^l)(\partial_l A^k)$$

(summation over l is understood). The first term is recognized as T^{0k} in Eq. (4.38). The second term is an addendum to the expression (4.38), according to

$$T^{0k} \rightarrow -(\partial^0 A_l)(\partial^k A^l) + (\partial^0 A_l)(\partial^l A^k) \tag{4.41}$$

$$= (\partial^0 A_l)(\partial^l A^k - \partial^k A^l) = F^0{}_l F^{lk}.$$

With this supplement, T^{0k} is cast into the form corresponding to the energy-momentum tensor

$$T^{\mu\nu} = F^\mu{}_\rho F^{\rho\nu} - \frac{1}{4}g^{\mu\nu}\left(F^{\rho\sigma} F_{\rho\sigma}\right)$$

kown from electrodynamics, as specified in Sec. 1.6. This form is gauge invariant. Instead, T^{0k} in (4.38) is not invariant under gauge transformations even within the class defining the radiation gauge, and a modification is necessary for a physically meaningful result. The augmentation performed above by the shift (4.41) utilizes an ambiguity of the energy-momentum tensor with respect to adding an extra term

$$T^{\mu\nu} \rightarrow T^{\mu\nu} + \partial_\rho\left(F^{\mu\rho} A^\nu\right) \tag{4.42}$$

which maintains the conservation law $\partial_\mu T^{\mu\nu} = 0$ because of

$$\partial_\mu \partial_\rho\left(F^{\mu\rho} A^\nu\right) = 0$$

resulting from the antisymmetry of $F^{\mu\nu}$. For the radiation field with $A^0 = 0$, the extra term in Eq. (4.42) for $\mu = 0$ turns into the second term in Eq. (4.41) as one can easily verify,

$$\partial_\rho\big(F^{0\rho}A^k\big) = F^{0\rho}\big(\partial_\rho A^k\big) = F^{0l}\big(\partial_l A^k\big) = F^0_{\;l}\big(\partial^l A^k\big) = \big(\partial^0 A_l\big)\big(\partial^l A^k\big)$$

since $\partial_\rho F^{0\rho} = 0$ holds as part of Maxwell's equations. Note that the energy density given by T^{00} is not affected by the augmentation because the extra term vanishes for $\mu = \nu = 0$.

4.2.2   *Rotational invariance and angular momentum*

Invariance under rotations leads to the conservation law of angular momentum. For a given field $\phi(x)$ (which can represent also a set of fields ϕ_j) the corresponding Lagrangian

$$\mathscr{L}(x) = \mathcal{L}\big(\phi(x), \partial_\mu\phi(x)\big) \tag{4.43}$$

depends on x only via the fields and their derivatives. A Lorentz transformation $x^\mu \to x'^\mu$ in terms of an infinitesimal rotation by an angle θ around an axis $\vec{n}$ can be specified in the following way,

$$x'^0 = x^0, \qquad \vec{x}' = \vec{x} - \delta\vec{x} \quad \text{with} \quad \delta\vec{x} = \theta\,\vec{n} \times \vec{x} = \vec{\theta} \times \vec{x}, \quad \vec{\theta} = \theta\vec{n}. \tag{4.44}$$

4.2.2.1   *Scalar field*

First we consider a single scalar field $\phi(x)$ describing particles without spin. The following steps are analogous to those in the derivation of the energy-momentum tensor. Invariance of the Lorentz scalar $\mathcal{L}$ means

$$\mathscr{L}'(x') = \mathscr{L}(x) = \mathscr{L}(x(x')) = \mathscr{L}(x' + \delta x). \tag{4.45}$$

Instead of x' we write x once again and expand up to the first order,

$$\mathscr{L}(x + \delta x) - \mathscr{L}(x) = \delta x^k\,\partial_k\mathscr{L}(x) = \partial_\mu\big(g^\mu_{\;k}\mathscr{L}\,\delta x^k\big). \tag{4.46}$$

For the last step the following identities were used:

$$\partial_\mu\big(g^\mu_{\;k}\mathscr{L}\,\delta x^k\big) = \delta x^k\,\partial_\mu\big(g^\mu_{\;k}\mathscr{L}\big) + g^\mu_{\;k}\mathscr{L}\,\partial_\mu\big(\delta x^k\big)$$

and

$$g^\mu_{\;k}\,\partial_\mu\big(\delta x^k\big) = g^\mu_{\;k}\,\partial_\mu\big(\epsilon^{klm}\theta_l x_m\big) = \theta_l\,\epsilon^{klm}\,g^\mu_{\;k}\big(\partial_\mu x_m\big) = -\theta_l\,\epsilon^{klm}\,g_{mk} = 0.$$

On the other hand, for the composed function (4.43) the following expansion holds,

$$\mathcal{L}(x+\epsilon) - \mathcal{L}(x) = \left[\frac{\partial \mathcal{L}}{\partial \phi}(\delta\phi) + \left(\frac{\partial \mathcal{L}}{\partial(\partial_\mu \phi)} \right) \delta(\partial_\mu \phi) \right]_{\phi=\phi(x)} \tag{4.47}$$

$$= \frac{\partial \mathcal{L}}{\partial \phi}(\delta\phi) + \left(\frac{\partial \mathcal{L}}{\partial(\partial_\mu \phi)} \right) \partial_\mu(\delta\phi)$$

$$= \frac{\partial \mathcal{L}}{\partial \phi}(\delta\phi) - \partial_\mu \left(\frac{\partial \mathcal{L}}{\partial(\partial_\mu \phi)} \right) (\delta\phi) + \partial_\mu \left(\frac{\partial \mathcal{L}}{\partial(\partial_\mu \phi)}(\delta\phi) \right)$$

$$= \partial_\mu \left(\frac{\partial \mathcal{L}}{\partial(\partial_\mu \phi)}(\delta\phi) \right)$$

because of the equation of motion for ϕ. $\delta\phi$ is the change of ϕ under the infinitesimal rotation (4.44),

$$\delta\phi = \phi(x+\delta x) - \phi(x) = \delta x^k \, \partial_k \phi(x) = \left(\vec{\theta} \times \vec{x} \right)^k \partial_k \phi \tag{4.48}$$

$$= \epsilon^{klm} \theta_l x_m \, \partial_k \phi = \theta_l \, \epsilon^{lmk} x_m \, \partial_k \phi = \theta^l \, \epsilon_{lm}{}^k x^m \, \partial_k \phi$$

$$= \vec{\theta} \cdot (\vec{x} \times \nabla)\phi = i\,\vec{\theta} \cdot \left(-i\vec{x} \times \nabla \right)\phi.$$

Combining Eqs. (4.46) and (4.47) yields

$$\partial_\mu \left(g^\mu{}_k \mathcal{L} \, \delta x^k - \frac{\partial \mathcal{L}}{\partial(\partial_\mu \phi)}(\delta\phi) \right) = 0$$

and by insertion of (4.48) one obtains

$$\partial_\mu \left(g^{\mu k} \mathcal{L} \, \epsilon_{klm} \theta^l x^m - i \frac{\partial \mathcal{L}}{\partial(\partial_\mu \phi)} \, \vec{\theta} \cdot \left(-i\vec{x} \times \nabla \right)\phi \right)$$

$$= \partial_\mu \left(g^{\mu k} \mathcal{L} \, \epsilon_{klm} x^m - i \frac{\partial \mathcal{L}}{\partial(\partial_\mu \phi)} \left(-i\vec{x} \times \nabla \right)^l \phi \right) \theta^l = 0.$$

Since θ^l are arbitrary and independent, the following identity holds for each l,

$$\partial_\mu \left(g^{\mu k} \mathcal{L} \, \epsilon_{klm} x^m - i \frac{\partial \mathcal{L}}{\partial(\partial_\mu \phi)} \left(-i\vec{x} \times \nabla \right)^l \phi \right) = 0. \tag{4.49}$$

This vanishing divergence of the angular-momentum tensor displays the conservation law of angular momentum. The $\mu=0$ component represents the angular-momentum density. By integration the conserved total angular momentum of the field is obtained,

$$\int d^3x \left(-i \frac{\partial \mathcal{L}}{\partial(\partial_0 \phi)} \left(-i \vec{x} \times \nabla \right) \phi \right) = \vec{L} \tag{4.50}$$

which for a scalar field is identical to the orbital angular momentum. The explicit Lagrangian (4.3) provides the derivative

$$\frac{\partial \mathcal{L}}{\partial(\partial_0 \phi)} = \partial^0 \phi = \partial_0 \phi$$

for insertion into Eq. (4.50), yielding

$$\vec{L} = \int d^3x \left(-i\partial_0 \phi \right) \left[\vec{x} \times (-i\nabla) \right] \phi \equiv \int d\vec{L} \tag{4.51}$$

for the angular momentum of the field. It is instructive to combine this expression with the momentum density (4.31). With

$$d\vec{P} = d^3x \left(-i\partial_0 \phi \right) \left(-i\nabla \phi \right)$$

one can write

$$d\vec{L} = \vec{x} \times d\vec{P},$$

which is the angular momentum of the volume element d^3x. Hence one finds the representation of the field angular momentum by

$$\vec{L} = \int \vec{x} \times d\vec{P} \tag{4.52}$$

which allows an easy interpretation of the formal-looking expression (4.51) for $\vec{L}$. In case of several scalar fields ϕ_j one has to extend Eq. (4.50), summing over ϕ_j.

4.2.2.2 *Dirac field*

The next application concerns particles with spin $1/2$. Starting point is the Dirac Lagrangian (4.6) with the Dirac field $\psi(x)$. An infinitesimal rotation (4.44) with $\vec{\theta} = \theta \vec{n}$ transforms the spinor ψ according to (see Sec. 2.2)

$$\psi(x) \rightarrow e^{\frac{i}{2}\vec{\theta}\cdot\vec{\Sigma}} \psi(x + \delta x) = \left(1 + \frac{i}{2}\vec{\theta}\cdot\vec{\Sigma} \right) \psi(x + \delta x)$$

$$= \psi(x + \delta x) + \frac{i}{2} \left(\vec{\theta}\cdot\vec{\Sigma} \right) \psi(x)$$

$$= \psi(x) + \delta\psi. \tag{4.53}$$

The part of $\delta\psi$ resulting exclusively from the substitution of the space–time coordinates is completely analogous to that of the scalar field specified in Eq. (4.48). Hence, the effect of the rotation is described by

$$\delta\psi = i\,\vec{\theta}\cdot(-i\vec{x}\times\nabla)\psi + \frac{i}{2}(\vec{\theta}\cdot\vec{\Sigma})\psi = i\,\vec{\theta}\cdot\left(-i\vec{x}\times\nabla\psi + \frac{1}{2}\vec{\Sigma}\psi\right) \qquad (4.54)$$

or equivalently written for the individual components,

$$\delta\psi_a = i\,\vec{\theta}\cdot\left(-i\vec{x}\times\nabla\psi_a + \frac{1}{2}\sum_b\vec{\Sigma}_{ab}\,\psi_b\right). \qquad (4.55)$$

New in comparison with the scalar field is the spin part of $\delta\psi$, represented by the second term. Accordingly, the integrand in Eq. (4.50) has to be extended by the spin part when transfered to the spinor field. Moreover, one has to take into account that summation over the spinor components is required. In this way one obtains the angular-momentum density of the Dirac field,

$$\vec{\mathcal{J}} = -i\sum_c \frac{\partial\mathcal{L}}{\partial(\partial_0\psi_c)}\left(-i\vec{x}\times\nabla\psi_c + \frac{1}{2}\sum_d\vec{\Sigma}_{cd}\,\psi_d\right). \qquad (4.56)$$

Inserting the derivatives (4.32) yields

$$\vec{\mathcal{J}} = \sum_{a,c}\overline{\psi}_a\gamma^0_{ac}\left(-i\vec{x}\times\nabla\psi_c + \frac{1}{2}\sum_d\vec{\Sigma}_{cd}\,\psi_d\right) \qquad (4.57)$$

$$= \psi^\dagger\left(-i\vec{x}\times\nabla\psi + \frac{1}{2}\vec{\Sigma}\,\psi\right),$$

and finally the total angular momentum appears as the sum

$$\vec{J} = \int d^3x\,\vec{\mathcal{J}} = \vec{L} + \vec{S} \qquad (4.58)$$

of orbital angular momentum and spin,

$$\vec{L} = \int d^3x\,\psi^\dagger[\vec{x}\times(-i\nabla)]\psi = \int \vec{x}\times d\vec{P}, \qquad \vec{S} = \int d^3x\,\psi^\dagger\left[\frac{1}{2}\Sigma\right]\psi. \qquad (4.59)$$

The expression for the spin operator was already encountered in Sec. 2.2.

4.2.2.3 *Electromagnetic radiation field*

A third example is the electromagnetic radiation field, with the Lagrangian (4.36) specified in radiation gauge. For describing infinitesimal rotations we need the generators of 3-dimensional rotations, represented by the matrices $\vec{\mathcal{S}} = (\mathcal{S}^1, \mathcal{S}^2, \mathcal{S}^3)$. They are explicity given by

$$\mathcal{S}^1 = \begin{pmatrix} 0 & 0 & 0 \\ 0 & 0 & -i \\ 0 & i & 0 \end{pmatrix}, \qquad \mathcal{S}^2 = \begin{pmatrix} 0 & 0 & i \\ 0 & 0 & 0 \\ -i & 0 & 0 \end{pmatrix}, \qquad \mathcal{S}^3 = \begin{pmatrix} 0 & -i & 0 \\ i & 0 & 0 \\ 0 & 0 & 0 \end{pmatrix}.$$

$$(4.60)$$

Under the infinitesimal rotation (4.44) with the transformation matrix expanded in terms of the generators,

$$R^k_l = \left[\mathbf{1} + i \left(\vec{\theta} \cdot \vec{\mathcal{S}} \right) \right]^k_l \tag{4.61}$$

the components of the vector potential transform according to

$$A^k(x) \rightarrow R^k_l \, A^l(x + \delta x) \tag{4.62}$$

$$= A^k(x + \delta x) + i \left(\vec{\theta} \cdot \vec{\mathcal{S}} \right)^k_l A^l(x)$$

$$= A^k(x) + \delta A^k.$$

The part of the variation originating from the substitution of the coordinates once more leads to the orbital angular momentum, in analogy to the scalar field and the Dirac field. The new feature is the spin part of δA^k leading to the spin operator. The spin contribution to the angular-momentum density is obtained by similar steps as in the Dirac case in Eq. (4.56), yielding

$$\vec{\mathcal{J}}_{\text{spin}} = -i \frac{\partial \mathcal{L}}{\partial (\partial_0 A^k)} \left(\vec{\mathcal{S}} \right)^k_l A^l. \tag{4.63}$$

Finally, the spin operator follows by integration,

$$\vec{S} = \int d^3x \, \vec{\mathcal{J}}_{\text{spin}} = \int d^3x \, (i \partial_0 A_k) \left(\vec{\mathcal{S}} \right)^k_l A^l. \tag{4.64}$$

4.3 Gauge Symmetry and QED

Symmetries under transformations that are not space–time transformations are denoted as *internal symmetries* of a physical system. Continuous internal symmetries, i.e. those depending on continuous parameters, imply conservation laws according to Noether's theorem, as in case of space–time symmetries.

As a specific system we consider free spin-1/2 particles with mass m described by the Dirac Lagrangian $\mathcal{L}$ given in Eq. (4.6). It is immediately clear that $\mathcal{L}$ is invariant under phase transformations

$$\psi \to e^{i\alpha}\,\psi, \qquad \alpha \in \mathbb{R}, \tag{4.65}$$

with an arbitrary real parameter α. Invariance of $\mathcal{L}$ under infinitesimal phase transformations

$$\psi \to (1 + i\delta\alpha)\,\psi = \psi + i\delta\alpha\,\psi = \psi + \delta\psi$$

means that the variation of $\mathcal{L}$ is zero,

$$\delta\mathcal{L} = 0 = \sum_a \left[\frac{\partial\mathcal{L}}{\partial\psi_a}\,\delta\psi_a + \frac{\partial\mathcal{L}}{\partial\overline{\psi}_a}\,\delta\overline{\psi}_a + \frac{\partial\mathcal{L}}{\partial(\partial_\mu\psi_a)}\,\partial_\mu(\delta\psi_a) \right]. \tag{4.66}$$

Making use of the equations of motion

$$\frac{\partial\mathcal{L}}{\partial\psi_a} = \partial_\mu\left(\frac{\partial\mathcal{L}}{\partial(\partial_\mu\psi_a)} \right), \qquad \frac{\partial\mathcal{L}}{\partial\overline{\psi}_a} = 0,$$

one obtains from Eq. (4.66)

$$\sum_a \partial_\mu\left(\frac{\partial\mathcal{L}}{\partial(\partial_\mu\psi_a)}\,\delta\psi_a \right) = \partial_\mu\left(i\sum_a \frac{\partial\mathcal{L}}{\partial(\partial_\mu\psi_a)}\,\psi_a \right)\delta\alpha$$

$$= \partial_\mu\left(-\underbrace{\sum_a \overline{\psi}_a\gamma^\mu_{ab}\,\psi_b}_{\overline{\psi}\gamma^\mu\psi} \right)\delta\alpha = 0.$$

Hence, because of $\delta\alpha \neq 0$, a conserved current $j^\mu = \overline{\psi}\gamma^\mu\psi$ exists, with $\partial_\mu j^\mu = 0$ and with a conserved charge

$$Q = \int d^3x\, j^0 = \int d^3x\, \overline{\psi}\gamma^0\psi,$$

as encountered already on earlier occasions. Now current conservation appears as a consequence of an abstract symmetry under phase transformations. This symmetry is denoted as *global gauge symmetry*, and the transformations (4.65) are called *global gauge transformations*. Mathematically they form an Abelian group, the unitary 1-parameter group $U(1)$.

4.3.1 *Local gauge transformations*

In Eq. (4.65) the phase α is a global parameter for the entire field $\psi(x)$. The choice $\alpha = \alpha(x)$ as a real function of x, instead, induces a local phase transformation

$$\psi(x) \to \psi'(x) = e^{i\alpha(x)}\,\psi(x) \qquad (4.67)$$

under which $\mathcal{L}$ is no longer invariant because of $\partial_\mu(e^{i\alpha}\psi) \neq e^{i\alpha}\,\partial_\mu\psi$.

Invariance under local phase transformations is achieved replacing the partial derivative in $\mathcal{L}$ by an appropriate *covariant derivative*,

$$\partial_\mu \to D_\mu = \partial_\mu + ieA_\mu \qquad (4.68)$$

corresponding to the minimal substitution. In this way a vector field A_μ is introduced which is also transformed, in combination with the phase transformation $\psi \to \psi'$ according to

$$\psi \to \psi' = e^{i\alpha}\,\psi, \qquad \alpha = \alpha(x) \qquad (4.69)$$

$$A_\mu \to A'_\mu = A_\mu - \frac{1}{e}\,\partial_\mu\alpha.$$

These combined transformations are denoted as *local gauge transformations*; they form the *local U(1)*, the gauge group of the electromagnetic interaction. The transformations of the vector field in (4.69) have a classical analogon in terms of the gauge transformations of the vector potential in classical electrodynamics.

As one can easily verify, the Lagrangian obtained by the substitution (4.68),

$$\mathcal{L} = \overline{\psi}\left(i\gamma^\mu D_\mu - m\right)\psi \qquad (4.70)$$

is invariant under local gauge transformations (4.69). The introduction of the vector field A_μ required for this formal symmetry has an immediate physical consequence: an interaction with the Dirac field via the coupling of A_μ to the conserved current,

$$\overline{\psi}\left(i\gamma^\mu D_\mu - m\right)\psi = \underbrace{\overline{\psi}\left(i\gamma^\mu \partial_\mu - m\right)\psi}_{\mathcal{L}_0}\ \underbrace{-\,e\,\overline{\psi}\gamma^\mu\psi\,A_\mu}_{\mathcal{L}_{\text{int}}} \qquad (4.71)$$

splitting $\mathcal{L}$ into $\mathcal{L}_0$ for the free Dirac field and an interaction term,

$$\mathcal{L} = \mathcal{L}_0 + \mathcal{L}_{\text{int}}, \qquad \mathcal{L}_{\text{int}} = -e\,j^\mu A_\mu. \qquad (4.72)$$

$\mathcal{L}_{\text{int}}$ describes the fundamental electromagnetic interaction of fermions if A_μ is the vector potential of the electromagnetic field. The purely formal parameter e in the covariant derivative (4.68) now appears as a coupling constant. However, A_μ in Eq. (4.71) so far can be at best a given external field since an important contribution does not exist: the vector potential needs a kinetic term $\mathcal{L}_A$ in the Lagrangian to become an autonomous dynamical field. In Sec. 4.1 the expression

$$\mathcal{L}_A = -\frac{1}{4}F_{\mu\nu}F^{\mu\nu} \tag{4.73}$$

was shown to reproduce the free Maxwell equations. $\mathcal{L}_A$ is invariant under gauge transformations on its own and thus can be added to $\mathcal{L}$ in Eq. (4.70) maintaining the overall gauge symmetry. In this way the Lagrangian of QED is obtained,

$$\boxed{\mathcal{L}_{\mathrm{QED}} = \overline{\psi}\left(i\gamma^\mu\partial_\mu - m\right)\psi - e\,\overline{\psi}\gamma^\mu\psi\,A_\mu - \frac{1}{4}F_{\mu\nu}F^{\mu\nu}} \tag{4.74}$$

accommodating both Lorentz and gauge symmetry. The A_μ field is massless, a mass term as in Eq. (4.12) is incompatible with local gauge symmetry. The equations of motion[1]

$$\partial_\mu\left(\frac{\partial\mathcal{L}}{\partial(\partial_\mu A_\nu)}\right) = \frac{\partial\mathcal{L}}{\partial A_\nu}, \qquad \frac{\partial\mathcal{L}}{\partial\overline{\psi}_a} = 0$$

follow as the inhomogenious Maxwell equations

$$\partial_\mu F^{\mu\nu} = e\,\overline{\psi}\gamma^\mu\psi = e\,j^\mu \tag{4.75}$$

and the Dirac equation with the electromagnetic field

$$\left(i\gamma^\mu\partial_\mu - m\right)\psi = e\,\gamma^\mu\psi\,A_\mu, \tag{4.76}$$

constituting the basic field equations of QED.

Hamilton operator. To establish the connection to Chap. 3 the Hamiltonian is needed for the perturbative treatment in the interaction picture. It is obtained as $H = \int d^3x\,\mathcal{H}$ from the density $\mathcal{H}$, which follows from the Lagrangian by a Legendre transformation. Generalization from particle mechanics to field theory yields the Legendre transformed of the QED

---

[1] For simplicity, the notation $\mathcal{L}$ is kept for $\mathcal{L}_{\mathrm{QED}}$.

Lagrangian as follows (with summation also over spinor indices),

$$\mathcal{H} = (\partial_0 \psi_a) \frac{\partial \mathcal{L}}{\partial(\partial_0 \psi_a)} + (\partial_0 A_k) \frac{\partial \mathcal{L}}{\partial(\partial_0 A_k)} - \mathcal{L} \qquad (4.77)$$

$$= \psi^\dagger \left[-i\vec{\alpha} \cdot \nabla + \beta m \right] \psi + (\partial_0 A_k) F^{k0} + \frac{1}{4} F_{\mu\nu} F^{\mu\nu} + e\, j^\mu A_\mu$$

$$= \mathcal{H}_0^{\text{Dirac}} + \vec{E}^2 - \frac{1}{2}(\vec{E}^2 - \vec{B}^2) + (\nabla A_0) \cdot \vec{E} + e\, j^\mu A_\mu$$

$$= \mathcal{H}_0^{\text{Dirac}} + \frac{1}{2}(\vec{E}^2 + \vec{B}^2) + (\nabla A_0) \cdot \vec{E} + e\, j^\mu A_\mu.$$

The first term is the Hamiltonian density for the free Dirac field and coincides with the T^{00} component of the energy–momentum tensor (4.33). The second term is the energy density of the electromagnetic field, but differently to the free radiation field in Eq. (4.37) also the Coulomb contribution is contained. The third term can be rearranged by means of $\nabla \cdot \vec{E} = \rho = e j^0$ as follows,[2]

$$(\nabla A_0) \cdot \vec{E} = \nabla \cdot (A_0 \vec{E}) - A_0 (\nabla \cdot \vec{E}) = -e\, j^0 A_0$$

and thus cancels the 0-component in the interaction part of Eq. (4.77). Hence, the Hamiltonian takes the form

$$H = \int d^3x\, \mathcal{H} = H_0^{\text{Dirac}} + \frac{1}{2} \int d^3x\, (\vec{E}^2 + \vec{B}^2) - e \int d^3x\, \vec{j} \cdot \vec{A} \qquad (4.78)$$

serving as the basis for perturbation theory and Feynman graphs (for the effective use of $\mathcal{H}_{\text{int}} = -\mathcal{L}_{\text{int}} = e\, j^\mu A_\mu$ see the discussion in Sec. 3.9).

4.3.2 *Summary*

QED is a highly successful theory and best confirmed by experiments. On the other hand, it arises from the symmetry principle of gauge invariance by means of simple rules and very few assumptions. This gives a strong motivation to consider QED as a guideline for constructing an appropriate theory also for the other fundamental interactions. Therefore, the essential steps leading to the structure of QED by the requirement of local gauge invariance are put together allowing for a generalization to other gauge groups.

---

[2]The total divergence yields an irrelevant surface term in the integral for H and can be omitted.

(i) Setting up a Lagrangian $\mathcal{L}_0$ for a physical system of free fermions with a global gauge symmetry, which means that $\mathcal{L}_0$ is invariant under a group of global gauge transformations.

(ii) Extension to local gauge symmetry by constructing a covariant derivative D_μ in such a way that the substitution $\partial_\mu \to D_\mu$ in $\mathcal{L}_0$ yields a Lagrangian $\mathcal{L}$ that is invariant under local gauge transformations. By this substitution an additional vector field A_μ is introduced and, moreover, an interaction between the fermions and A_μ is induced.

(iii) Promotion of A_μ to a dynamical field by adding a kinetic term $\mathcal{L}_A$ that is invariant under local gauge transformations as well. The dynamical field A_μ as a quantum field belongs to spin-1 particles (vector bosons) with mass zero.

The resulting Lagrangian $\mathcal{L} + \mathcal{L}_A$ describes a system consisting of fermions and massless vector bosons with a mutual interaction in terms of a current–field coupling $\sim j^\mu A_\mu$. The current is the conserved Noether current following from the global gauge symmetry under (i).

Terminology. The vector field A_μ is denoted as *gauge field*, the associated massless spin-1 particles are referred to as *gauge bosons*.

A special feature of QED is the Abelian character of the gauge group $U(1)$. To transfer the construction principle of a gauge-invariant field theory to the strong and weak interactions necessitates generalization of the concept of gauge transformations to non-Abelian groups.

4.4   Non-Abelian Gauge Symmetries

A generalization of the phase transformations (4.65) to transformations that do not commute,

$$\psi \to U\psi \qquad \text{with} \quad U_2 U_1 \psi \neq U_1 U_2 \psi$$

requires matrices and hence a certain multiplicity of ψ,

$$\psi = \begin{pmatrix} \psi_1 \\ \cdot \\ \cdot \\ \cdot \\ \psi_n \end{pmatrix}, \qquad \overline{\psi} = (\overline{\psi}_1, \dots \overline{\psi}_n). \tag{4.79}$$

Each ψ_k of the multiplet is a 4-component Dirac spinor, and $\overline{\psi}_k$ the corresponding adjoint spinor; as a quantum field ψ_k describes fermions/antifermions of a given species. The n spinor fields of the multiplet refer to a system of n fermion species that are treated initially as free particles. For the construction of a gauge invariant theory with interaction we proceed according to the steps (i)–(iii) formulated in the previous section.

4.4.1 *Global gauge symmetry*

The Lagrangian for the system of free particles is the sum of the individual Dirac parts. In case of equal masses for all fermions k the Lagrangian can be written in a compact way in terms of the multiplets (4.79),

$$\mathcal{L}_0 = \sum_{k=1}^{n} \overline{\psi}_k \big(i\gamma^\mu \partial_\mu - m\big)\psi_k = \overline{\psi}\big(i\gamma^\mu \partial_\mu - m\big)\psi. \tag{4.80}$$

$\mathcal{L}_0$ is invariant under transformations of the type

$$\psi \to \psi' = U\psi \tag{4.81}$$

if U is a unitary and x-independent $n \times n$ matrix, since for $U^\dagger = U^{-1}$ the following relations hold,

$$\overline{\psi}'\psi' = \overline{\psi}\, U^\dagger U \psi = \overline{\psi}\psi,$$

$$\overline{\psi}'\partial_\mu \psi' = \overline{\psi}\, U^\dagger \partial_\mu U \psi = \overline{\psi}\partial_\mu \psi.$$

The transformations (4.81) are generalizations of the phase transformations (4.65) and thus depict the *global gauge transformations*. They form a group, denoted as the gauge group accordingly.

Particularly relevant for particle physics is the group of unitary $n \times n$ matrices with determinant $\det = +1$ (special unitary group),

$$\boxed{\text{group } SU(n)}$$

Each element U of this group contains n^2 complex entries restricted by n^2 unitarity conditions and the condition $\det = 1$. Hence,

$$2n^2 - (n^2 + 1) = n^2 - 1 = N \tag{4.82}$$

independent real parameters labeled as $\alpha_1, \ldots \alpha_N$ remain for specifying a group element U,

$$U = U(\alpha_1, \ldots \alpha_N).$$

For infinitesimal parameters $\delta\alpha_1, \ldots \delta\alpha_N$, U can be expanded to first order

$$U = \mathbf{1} + i \sum_{a=1}^{N} T_a \, \delta\alpha_a \equiv \mathbf{1} + i \, T_a \, \delta\alpha_a \tag{4.83}$$

(summation convention also applied to the parameter indices). The quantities T_a are $n \times n$ matrices, they are denoted as the *generators* of the gauge group. Unitarity of U and $\det U = 1$ require that they are Hermitian with vanishing trace,

$$T_a^\dagger = T_a, \qquad \mathrm{Tr}(T_a) = 0. \tag{4.84}$$

The generators fulfill the commutation relations

$$\left[T_a, T_b\right] = i f_{abc} \, T_c \tag{4.85}$$

with real numbers f_{abc}, the *structure constants* of $SU(n)$. They are totally antisymmetric in the indices,

$$f_{abc} = -f_{bac} = -f_{acb} = \cdots$$

The normalization of T_a is defined by convention, usually in terms of the condition

$$\mathrm{Tr}\,(T_a T_b) = \frac{1}{2}\,\delta_{ab}. \tag{4.86}$$

With the help of the generators, the group elements $U(\alpha_1, \ldots \alpha_N)$ for finite parameters are represented as the exponentiated version of the expansion (4.83),

$$U = \mathbf{1} + i\,\alpha_a\,T_a + \cdots = e^{i\,\alpha_a T_a}. \tag{4.87}$$

Terminology. A group with elements depending on N real continuous parameters is denoted as a N-dimensional *Lie group*; the generators with the commutation relations (4.85) form the associated *Lie algebra*.

Example 1: $SU(2)$

The unitary 2×2 matrices with $\det = +1$ are specified by $N = 3$ real parameters $\alpha_1, \alpha_2, \alpha_3$. The generators

$$T_a = \frac{1}{2}\,\sigma_a, \qquad a = 1, 2, 3 \tag{4.88}$$

can be expressed in terms of the Pauli matrices $\sigma_{1,2,3}$. They obey the normalization (4.86) and the commutation relations

$$\left[T_a, T_b\right] = i\,\epsilon_{abc}\,T_c \tag{4.89}$$

which are formally identical to those of angular momentum. The Lie algebra of $SU(2)$ thus carries the name *isospin* algebra and the group is the *isospin* group. In physics applications it is part of the gauge group of the weak interaction.

Example 2: $SU(3)$

The unitary 3×3 matrices with det $= +1$ are specified by $N = 9 - 1 = 8$ real parameters $\alpha_1, \ldots \alpha_8$, related to 8 generators. The generators form the 8-dimensional Lie algebra with the specific structure constants f_{abc} of $SU(3)$ listed in Table 4.1. Conventionally the generators T_a are expressed in terms of the Gell-Mann matrices $\lambda_1, \ldots \lambda_8$,

$$T_a = \frac{1}{2}\lambda_a, \qquad a = 1, \ldots 8 \tag{4.90}$$

given by

$$\lambda_1 = \begin{pmatrix} 0 & 1 & 0 \\ 1 & 0 & 0 \\ 0 & 0 & 0 \end{pmatrix}, \quad \lambda_2 = \begin{pmatrix} 0 & -i & 0 \\ i & 0 & 0 \\ 0 & 0 & 0 \end{pmatrix}, \quad \lambda_3 = \begin{pmatrix} 1 & 0 & 0 \\ 0 & -1 & 0 \\ 0 & 0 & 0 \end{pmatrix},$$

$$\lambda_4 = \begin{pmatrix} 0 & 0 & 1 \\ 0 & 0 & 0 \\ 1 & 0 & 0 \end{pmatrix}, \quad \lambda_5 = \begin{pmatrix} 0 & 0 & -i \\ 0 & 0 & 0 \\ i & 0 & 0 \end{pmatrix},$$

$$\lambda_6 = \begin{pmatrix} 0 & 0 & 0 \\ 0 & 0 & 1 \\ 0 & 1 & 0 \end{pmatrix}, \quad \lambda_7 = \begin{pmatrix} 0 & 0 & 0 \\ 0 & 0 & -i \\ 0 & i & 0 \end{pmatrix}, \quad \lambda_8 = \frac{1}{\sqrt{3}}\begin{pmatrix} 1 & 0 & 0 \\ 0 & 1 & 0 \\ 0 & 0 & -2 \end{pmatrix}.$$

Historically, $SU(3)$ was introduced by Gell-Mann as the symmetry group for the (approximative) flavour symmetry of the hadron spectrum [GellMann

Table 4.1 Structure constants f_{abc} of $SU(3)$. All the other f_{abc} not available by permutation of the indices are zero.

abc	f_{abc}	abc	f_{abc}
123	1	345	$\frac{1}{2}$
147	$\frac{1}{2}$	367	$-\frac{1}{2}$
156	$-\frac{1}{2}$	458	$\frac{1}{2}\sqrt{3}$
246	$\frac{1}{2}$	678	$\frac{1}{2}\sqrt{3}$
257	$\frac{1}{2}$		

(1962)], being represented by the quark triplet u, d, s [GellMann (1964)]. In later course, Fritzsch and Gell-Mann postulated $SU(3)$ as the group for the colour symmetry of the quarks [Fritzsch (1972)] and thus as the gauge group of the strong interaction.

4.4.1.1 *Mathematical insertion: Representation*

- Each family of N Hermitian $d \times d$ matrices that obey the commutation rules (4.85) is a d-dimensional *representation of the Lie algebra* of $SU(n)$. For $d = n$ the defining representation of $SU(n)$ is obtained as the lowest-dimensional non-trivial representation, named *fundamental representation*. $d = N$ defines the *adjoint representation*.

  Example 1: The spin matrices for $s = \frac{1}{2}, 1, \frac{3}{2}, \ldots$ form the 2-, 3-, 4-,... dimensional representations of the $SU(2)$ Lie algebra. For $s = \frac{1}{2}$ one obtains the isospin generators (4.88) in the fundamental representation.

  Example 2: The generators (4.90) with the Gell-Mann matrices form the fundamental representation of the $SU(3)$ Lie algebra. Higher-dimensional representations exist as 6-, 8-, 10-, ... dimensional matrices.

- The group elements
  $$U(\alpha_1, \ldots \alpha_N) = e^{i\,\alpha_a T_a} \qquad \text{with} \quad T_a \Subset d-\text{dim. representation}$$
  form a d-dimensional *representation of the group $SU(n)$*. The group members U act as $d \times d$ matrices on d-dimensional multiplets Ψ,
  $$\Psi \;\rightarrow\; \Psi' = U\Psi.$$
  Der d-dimensional linear space of the Ψ is denoted as the representation space.

The fundamental representation in each case forms the basis in applications towards describing the symmetries of weak and strong interactions: the isospin doublets of leptons and quarks
$$\begin{pmatrix} \nu_e \\ e \end{pmatrix}, \begin{pmatrix} \nu_\mu \\ \mu \end{pmatrix}, \begin{pmatrix} \nu_\tau \\ \tau \end{pmatrix} \qquad \begin{pmatrix} u \\ d \end{pmatrix}, \begin{pmatrix} c \\ s \end{pmatrix}, \begin{pmatrix} t \\ b \end{pmatrix}$$
for the weak interaction, and the colour triplets of quarks (for each $q = u, d, s, c, b, t$)
$$\begin{pmatrix} q_1 \\ q_2 \\ q_3 \end{pmatrix}$$
for the strong interaction.

4.4.2 *Local gauge symmetry*

In the second step the global gauge transformations $U(\alpha_1, \ldots \alpha_N)$ with constant parameters are extended to local transformations by x-dependent functions $\alpha_a(x)$,

$$U\big(\alpha_1(x), \ldots \alpha_N(x)\big) \equiv U(x) = e^{i\alpha_a(x)T_a}, \qquad (4.91)$$

transforming the multiplet (4.79) according to

$$\psi(x) \to \psi'(x) = U(x)\,\psi(x). \qquad (4.92)$$

Invariance of the Lagrangian is obtained replacing the partial derivative by a suitable covariant derivative D_μ that is transformed either, $D_\mu \to D'_\mu$. It is decisive that the following condition is fulfilled as an operator identity,

$$U D_\mu = D'_\mu U, \qquad (4.93)$$

because this implies

$$\overline{\psi'}\,i\gamma^\mu D'_\mu \psi' = \overline{\psi}\,U^\dagger i\gamma^\mu D'_\mu U\psi = \overline{\psi}\,U^{-1}i\gamma^\mu\,U D_\mu\psi = \overline{\psi}\,i\gamma^\mu D_\mu\psi$$

and thus invariance of

$$\overline{\psi}\big(i\gamma^\mu D_\mu - m\big)\psi.$$

In order to fufill the condition (4.93) an ansatz is made for the minimal substitution, following QED,

$$D_\mu = \partial_\mu - ig\mathbf{W}_\mu \qquad (4.94)$$

where $\mathbf{W}_\mu$ is a $n \times n$ matrix because of the multiplicity of ψ. It can be written as a linear combination of the generators,

$$\mathbf{W}_\mu(x) = T_a\,W^a_\mu(x), \qquad (4.95)$$

with N vector fields W^a_μ, the *gauge fields*, as coefficients. g is for the moment an arbitrary constant. The condition (4.93) is fulfilled if and only if D_μ transforms according to

$$D'_\mu = \partial_\mu - ig\mathbf{W}'_\mu \qquad (4.96)$$

with

$$\mathbf{W}'_\mu = U\,\mathbf{W}_\mu\,U^{-1} - \frac{i}{g}\,\big(\partial_\mu U\big)\,U^{-1}. \qquad (4.97)$$

The transformations (4.92) and (4.97) in combination constitute the non-Abelian local gauge transformations and thus the generalization of (4.69).

In case of an infinitesimal gauge transformation

$$U = 1 + i\,\alpha_a T_a \qquad (4.98)$$

one obtains from Eq. (4.97) together with $\mathbf{W}'_\mu = T_a\,W'^a_\mu$ the transformation of the vector fields deviating from the Abelian case by the second term,

$$W'^a_\mu = W^a_\mu + f_{abc}\,W^b_\mu\,\alpha_c + \frac{1}{g}\partial_\mu\alpha_a. \qquad (4.99)$$

The substitution of the covariant derivative induces an interaction between the fermion fields and the gauge fields,

$$\overline{\psi}\big(i\gamma^\mu D_\mu - m\big)\psi = \mathcal{L}_0 + g\,\overline{\psi}\gamma^\mu T_a\psi\,W^a_\mu = \mathcal{L}_0 + \mathcal{L}_{\text{int}} \qquad (4.100)$$

where the interaction Lagrangian

$$\mathcal{L}_{\text{int}} = g\,j^\mu_a\,W^a_\mu \qquad (4.101)$$

describes the coupling of the gauge fields to N currents

$$j^\mu_a = \overline{\psi}\,\gamma^\mu T_a\psi = \overline{\psi}_k\,\gamma^\mu (T_a)_{kl}\,\psi_l. \qquad (4.102)$$

The generators can be considered the non-Abelian generalization of the electromagnetic charge Q, cf. the electromagnetic current in Eq. (3.60). The constant g in the covariant derivative now turns out to be a universal coupling constant.

4.4.2.1 *Amendment*

The concept of constructing a gauge invariant field theory has been exemplified for the specific case of a fermion multiplet. The same steps can be performed in complete analogy for a multiplet of scalar fields

$$\Phi = \begin{pmatrix}\phi_1 \\ \vdots \\ \phi_n\end{pmatrix}, \qquad \Phi^\dagger = \big(\phi_1^\dagger, \ldots \phi_n^\dagger\big) \qquad (4.103)$$

starting from a Lagrangian for n species of free spin-0 particles with equal mass,

$$\mathcal{L}_0 = \sum_{k=1}^{n}\big[(\partial_\mu\phi_k)^\dagger(\partial^\mu\phi_k) - m^2\phi_k^\dagger\phi_k\big] = (\partial_\mu\Phi)^\dagger(\partial^\mu\Phi) - m^2\,\Phi^\dagger\Phi, \qquad (4.104)$$

which shows the same symmetry under global $SU(n)$ transformations

$$\Phi \rightarrow \Phi' = U\Phi \qquad (4.105)$$

as the Dirac Lagrangian (4.80). Replacing ∂_μ by the covariant derivative (4.94) generates invariance under local gauge transformations as well and induces an interaction term between the scalar fields and the gauge fields,

$$\left(D_\mu \Phi\right)^\dagger \left(D^\mu \Phi\right) - m^2\, \Phi^\dagger \Phi = \mathcal{L}_0 + \mathcal{L}_{\text{int}}, \tag{4.106}$$

$$\mathcal{L}_{\text{int}} = g\left[\Phi^\dagger T_a(i\partial^\mu \Phi) - (i\partial^\mu \Phi^\dagger)T_a \Phi\right]W_\mu^a$$
$$+\, g^2\, g^{\mu\nu}\left(\Phi^\dagger T_a T_b \Phi\right)W_\mu^a\, W_\nu^b.$$

A new feature is the appearance of a 4-point coupling between two scalars and two vector fields.

An important physics application is found in the Higgs mechanism of the electroweak interaction where a $SU(2)$ doublet of scalar fields is coupled to the vector fields in the way described above (see Sec. 6.4).

4.4.3 *Dynamics of the gauge fields*

The last step towards completion of a gauge invariant field theory consists in adding a kinetic term $\mathcal{L}_W$ for the vector fields W_μ^a yielding the equations of motion and thus defining the dynamics of the gauge fields. The naive procedure in analogy to QED, namely replacing A_μ by W_μ^a and summing over a, is not constructive because the result is not invariant under local gauge transformations.

The gauge fields enter the covariant derivative (4.94) in terms of a matrix $\mathbf{W}_\mu$. Therefore, it is obvious to make an ansatz for the field strength tensor as a matrix either, with the replacement $\partial_\mu \to D_\mu$,

$$\mathbf{F}_{\mu\nu} = D_\mu \mathbf{W}_\nu - D_\nu \mathbf{W}_\mu \tag{4.107}$$

$$= \partial_\mu \mathbf{W}_\nu - \partial_\nu \mathbf{W}_\mu - ig\left[\mathbf{W}_\mu, \mathbf{W}_\nu\right].$$

The expansion in terms of the generators

$$\mathbf{F}_{\mu\nu} = T_a\, F_{\mu\nu}^a \tag{4.108}$$

contains N tensor fields as coefficients which by means of $F_{\mu\nu}^a = 2\operatorname{Tr}(\mathbf{F}_{\mu\nu}T_a)$ can be extracted from (4.107), reading

$$F_{\mu\nu}^a = \partial_\mu W_\nu^a - \partial_\nu W_\mu^a + g f_{abc}\, W_\mu^b\, W_\nu^c. \tag{4.109}$$

As one can easily verify, $\mathbf{F}_{\mu\nu}$ can be expressed as a commutator of two covariant derivatives,

$$\mathbf{F}_{\mu\nu} = \frac{i}{g}\left[D_\mu, D_\nu\right]. \tag{4.110}$$

This implies, by use of $D'_\mu = U D_\mu U^{-1}$, the following behaviour under gauge transformations,

$$\mathbf{F}'_{\mu\nu} = U\,\mathbf{F}_{\mu\nu}U^{-1}. \tag{4.111}$$

Consequently, a gauge invariant quantity is obtained by squaring and subsequent trace calculation,

$$\mathrm{Tr}\left(\mathbf{F}'_{\mu\nu}\mathbf{F}'^{\,\mu\nu}\right) = \mathrm{Tr}\left(U\,\mathbf{F}_{\mu\nu}\mathbf{F}^{\mu\nu}\,U^{-1}\right) = \mathrm{Tr}\left(\mathbf{F}_{\mu\nu}\mathbf{F}^{\mu\nu}\right). \tag{4.112}$$

Hence, a suitable Lagrangian for the gauge fields is given by

$$\mathcal{L}_W = -\frac{1}{2}\,\mathrm{Tr}\left(\mathbf{F}_{\mu\nu}\mathbf{F}^{\mu\nu}\right) = -\frac{1}{4}\,F_{\mu\nu}^a\,F^{a,\mu\nu}. \tag{4.113}$$

The quadratic part of $\mathcal{L}_W$ is responsible for the free propagation of the W-fields, but there are also trilinear and quadrilinear terms describing self interactions of the vector fields,

$$\mathcal{L}_W = -\frac{1}{4}\left(\partial_\mu W_\nu^a - \partial_\nu W_\mu^a\right)\left(\partial^\mu W^{a,\nu} - \partial^\nu W^{a,\mu}\right) \tag{4.114}$$

$$-\frac{g}{2}\,f_{abc}\left(\partial_\mu W_\nu^a - \partial_\nu W_\mu^a\right)W^{b,\mu}\,W^{c,\nu}$$

$$-\frac{g^2}{4}\,f_{abc}f_{ade}\,W_\mu^b\,W_\nu^c\,W^{d,\mu}\,W^{e,\nu}.$$

The self couplings are characteristic for non-Abelian gauge fields. Their structure is completely determined by gauge symmetry, and their strength is normalized by the universal coupling constant g. From a physics point of view, the self interaction means that the gauge fields themselves carry non-Abelian charges and thus on their part act as sources for the W_μ^a fields, besides the fermionic currents (4.102).

The total Lagrangian of the system consisting of fermion fields and gauge fields finally is given by

$$\mathcal{L} = \overline{\psi}\left(i\gamma^\mu D_\mu - m\right)\psi + \mathcal{L}_W. \tag{4.115}$$

The existence of conserved currents is a consequence of the symmetry of $\mathcal{L}$ under global gauge transformations. For infinitesimal, x-independent $\delta\alpha_a$

they read as follows,

$$\psi' = \psi + \delta\psi, \qquad\qquad \delta\psi = i\,T_a\psi\,\delta\alpha_a, \qquad (4.116)$$

$$W_\nu'^{\,a} = W_\nu^a + \delta W_\nu^a, \qquad \delta W_\nu^a = f_{abc}\,W_\nu^b\,\delta\alpha_c.$$

Invariance of $\mathcal{L}$ implies

$$\delta\mathcal{L} = \partial_\mu\left(\frac{\partial\mathcal{L}}{\partial(\partial_\mu\psi_{k\alpha})}\,\delta\psi_{k\alpha} + \frac{\partial\mathcal{L}}{\partial(\partial_\mu W_\nu^a)}\,\delta W_\nu^a\right) = 0$$

as one can derive using the equations of motion and performing the same steps as at the beginning of Sec. 4.3. With Eq. (4.116) and the derivatives

$$\frac{\partial\mathcal{L}}{\partial(\partial_\mu\psi_{k\alpha})} = \overline{\psi}_{k\beta}i\gamma^\mu_{\beta\alpha}, \qquad \frac{\partial\mathcal{L}}{\partial(\partial_\mu W_\nu^a)} = -F^{a,\mu\nu}$$

one obtains

$$\partial_\mu\left(-\overline{\psi}_{k\beta}\,\gamma^\mu_{\beta\alpha}(T_a)_{kl}\,\psi_{l\alpha} - f_{abc}\,F^{b,\mu\nu}W_\nu^c\right)\delta\alpha_a = -\left(\partial_\mu J_a^\mu\right)\delta\alpha_a = 0.$$

Since all $\delta\alpha_a$ are independent and arbitrary, current conservation $\partial_\mu J_a^\mu = 0$ follows for each of the N currents

$$J_a^\mu = \overline{\psi}\gamma^\mu T_a\psi + f_{abc}\,F^{b,\mu\nu}W_\nu^c = j_a^\mu + f_{abc}\,F^{b,\mu\nu}\,W_\nu^c. \qquad (4.117)$$

The fermionic currents j_a^μ are thus not separately conserved since the vector fields contribute to the conserved total current as well, due to the gauge charges on their own. The Euler-Lagrange equations of motion for the gauge fields,

$$\partial_\mu F^{a,\mu\nu} = -g\,J_a^\nu, \qquad (4.118)$$

are the generalized Maxwell equations with the conserved currents (4.117).

4.4.3.1 *Gauge field propagators*

The gauge fields W_a^μ as quantum fields describe N massless spin-1 particles, the gauge bosons, which in each case carry one of the N different gauge charges. The part of $\mathcal{L}_W$ quadratic in the fields determines the free dynamics and yields via Euler–Lagrange the free field equations for the gauge fields, according to Eq. (4.118) for vanishing coupling constant $g \to 0$,

$$\partial_\mu F^{a,\mu\nu} = \partial_\mu\left(\partial^\mu W^{a,\nu} - \partial^\nu W^{a,\mu}\right) \qquad (4.119)$$

$$= \Box W^{a,\nu} - \partial^\nu\left(\partial_\mu W^{a,\mu}\right) = 0.$$

Due to gauge freedom the W fields are not uniquely defined, and additional restricting conditions are required. A frequently imposed condition defines a gauge that is equivalent to the Lorentz gauge in electrodynamics, in the context here denoted as *Feynman gauge*,

$$\partial_\mu W^{a,\mu} = 0, \qquad a = 1, \ldots N. \tag{4.120}$$

In this class, each of the gauge fields obeys the homogeneous wave equation

$$\Box W^{a,\nu} = 0, \qquad a = 1, \ldots N. \tag{4.121}$$

The solutions can be represented by the complete orthogonal system of plane waves, corresponding to momentum and helicity eigenstates of the gauge bosons. As for photons, only two physical transverse polarizations exist corresponding to helicities ± 1 of massless particles.

The Green functions $D^{ab}_{\mu\nu}(x - y)$ of the gauge fields in Feynman gauge are determined as solutions of the inhomogeneous wave equation with a pointlike source,

$$\Box_{(x)} D^{ab}_{\mu\nu}(x - y) = \delta_{ab}\, g_{\mu\nu}\, \delta^4(x - y) \tag{4.122}$$

together with causal boundary conditions. In momentum space one obtains a N-fold copy of the photon propagator (3.28),

$$D^{ab}_{\mu\nu}(Q) = \frac{-g_{\mu\nu}}{Q^2 + i\varepsilon}\, \delta_{ab} \tag{4.123}$$

represented by a graphical symbol like that for the photon propagator in QED,

$$i\, D^{ab}_{\mu\nu}(Q) \qquad \underset{a,\mu \quad\; Q \quad\; b,\nu}{\sim\!\sim\!\sim\!\sim\!\sim\!\sim}$$

At the mathematial side, setting up the field equations for the gauge fields in a given gauge corresponds to a variational problem with side conditions,

$$\delta \int d^4 x\, \mathcal{L} = 0 \qquad \text{with} \quad \partial_\mu W^{a,\mu} = 0.$$

Equivalent is the following variational problem without side conditions but with an augmented Lagrangian,

$$\delta \int d^4 x \left(\mathcal{L} - \sum_a \frac{1}{2\xi_a} \left(\partial_\mu W^{a,\mu} \right)^2 \right) = 0 \tag{4.124}$$

including arbitrary parameters $\xi_a \in \mathbb{R}$. The resulting Euler–Lagrange equations for the gauge fields in the limit of vanishing coupling constant $g \to 0$

then take the form

$$\Box W^{a,\nu} - \left(1 - \frac{1}{\xi_a}\right) \partial^\mu \partial^\nu W^a_\mu = 0 \qquad (4.125)$$

For $\xi_a = 1$ the wave equation (4.121) in Feynman gauge is recovered.

The differential equation for the Green functions of the modified wave equation (4.125) is given by

$$\left(g^{\mu\rho}\Box - \left(1 - \frac{1}{\xi_a}\right)\partial^\mu\partial^\rho\right) D^{ab}_{\rho\nu}(x - y) = \delta_{ab}\, g^\mu_\nu\, \delta^4(x - y). \qquad (4.126)$$

Fourier transformation to momentum space yields the algebraic equation

$$\left(-Q^2 g^{\mu\rho} + \left(1 - \frac{1}{\xi_a}\right) Q^\mu Q^\rho\right) D^{ab}_{\rho\nu}(Q) = \delta_{ab}\, g^\mu_\nu \qquad (4.127)$$

which has the solution

$$D^{ab}_{\rho\nu}(Q) = \frac{\delta_{ab}}{Q^2 + i\varepsilon}\left(-g_{\rho\nu} + (1 - \xi_a)\frac{Q_\rho Q_\nu}{Q^2}\right). \qquad (4.128)$$

For $\xi_a = 1$ the propagators in Feynman gauge are recovered.

Due to the simple form of the propagators, the Feynman gauge is particularly suitable for applications and thus is frequently used in practical calculations. One has to take into consideration, however, that the respective propagators as 2-point functions in the framework of quantum field theory

$$D^{ab}_{\mu\nu}(x - y) = \langle 0 |\, T\, W^a_\mu(x)\, W^b_\nu(y)\, | 0 \rangle = \int \frac{d^4Q}{(2\pi)^4}\left(\frac{-ig_{\mu\nu}\,\delta_{ab}}{Q^2 + i\varepsilon}\right) e^{-iQ(x-y)}$$
$$\qquad (4.129)$$

contain not only the two transverse polarizations but also two unphysical polarizations of the free fields adding up in the polarization sum to $-g_{\mu\nu}$. One of these redundant polarizations becomes the mediator of the Coulomb force when the interaction is turned on so that three of the four degrees of freeedom have to be considered physical, according to the restriction of the four vector components by the gauge condition. Employing the Coulomb gauge $\partial_k W^{a,k} = 0$ instead of the Feynman gauge would make this connection more explicit and the physical interpretation more transparent since unphysical degrees of freedom are avoided; only the transverse polarizations propagate in the two-point functions and the Coulomb interaction is treated seperately, as illustrated for QED in Sec. 3.9. In the non-Abelian case, however, the Coulomb gauge proves rather unwieldy owing to the non-Abelian interaction terms and thus less suitable for practical applications. The same holds for other gauges with only physical degrees of freedom, as e.g. the axial gauge $n^\mu W_\mu^a = 0$ with a given constant 4-vector n^μ.

The advantage of a covariant gauge like Feynman gauge has its price by the presence of unphysical polarizations in the propagators. In QED they are irrelevant because of the conserved electromagnetic current or the global $U(1)$ symmetry, respectively. In non-Abelian gauge theories the conserved current (4.117) has a more complex structure owing to the self-couplings of the gauge fields. Accordingly, the spurious terms in the propagators generate non-vanishing contributions to S-matrix elements that are unphysical and gauge dependent. As far as the lowest order in the perturbative expansion is concerned, such contributions do not occur when for external gauge bosons transverse polarization vectors and the corresponding transverse polarization sum are used. At higher orders, however, with closed loops, the unphysical contributions are unavoidable whenever trilinear self couplings are present. Hence, it is necessary to subtract these contributions.

A systematic and elegant method consists in the augmentation of the Lagrangian by an additional term involving unphysical fields with appropriate couplings to the gauge fields in such a way that the unphysical components of the W-propagators in the loops are canceled. This method shall be illustrated in the next section.

4.4.4   *Gauge fixing and ghost fields*

The extra term in the Lagrangian (4.124) serves for defining a specific gauge and thus is denoted as a *gauge-fixing term* $\mathcal{L}_{\text{fix}}$. The real parameters ξ_a are called *gauge parameters* (often they are chosen uniform, $\xi_a = \xi$ for all $a = 1, \ldots N$). For the choice of gauge fixing like in (4.124) yielding the propagators (4.128) the name R_ξ-gauge is common. Note that without the ξ-term in Eq. (4.127) no solution for the propagators exists since the matrix on the left side has a vanishing determinant owing to the eigenvector Q_ρ with eigenvalue zero. This is a direct consequence of gauge invariance, which is broken by $\mathcal{L}_{\text{fix}}$. Accordingly, the propagators are gauge dependent.

The augmented Lagrangian $\mathcal{L} + \mathcal{L}_{\text{fix}}$ is no longer invariant under local gauge transformations (4.99). With the aim to restore the symmetry in an appropriate form one introduces a set of auxiliary scalar fields, u_a and $\overline{u}_a$ for $a = 1, \ldots N$, named *ghost fields* to indicate their unphysical character; often they are also called Faddeev–Popov ghosts according to the founders of the concept [Faddeev (1967)]. They are installed as dynamical fields by adding one more extra term to the Lagrangian,

$$\mathcal{L}_{\text{ghost}} = \left(\partial^\mu \, \overline{u}_a \right) \left(D_\mu^{\text{ad}} \right)_{ab} u_b \tag{4.130}$$

where the covariant derivative appears in the adjoint representation of the Lie algebra,

$$\left(D_\mu^{\text{ad}} \right)_{ab} = \delta_{ab} \, \partial_\mu - i \, g \left(T_c^{\text{ad}} \right)_{ab} W_\mu^c. \tag{4.131}$$

The ajoint representation is N-dimensional and the generators are represented by matrices with the structure constants as entries,

$$\left(T_c^{\text{ad}} \right)_{ab} = -i f_{cab}, \tag{4.132}$$

obeying the Lie-Algebra (4.85). For the proof the *Jacobi identity* is needed,

$$f_{ade}\, f_{bcd} + f_{bde}\, f_{cad} + f_{cde}\, f_{abd} = 0. \tag{4.133}$$

With the representation (4.132) the ghost Lagrangian gets the explicit form

$$\mathcal{L}_{\text{ghost}} = \left(\partial^\mu \, \overline{u}_a \right) \left(\partial_\mu u_a \right) + g \, f_{abc} \left(\partial^\mu \, \overline{u}_a \right) W_\mu^b \, u_c. \tag{4.134}$$

From the free kinetic part one obtains the propagators of the N scalar fields,

$$i \, D_{ab}(x - y) = \int \frac{d^4 Q}{(2\pi)^4} \, i \, D_{ab}(Q) \, e^{-iQ(x-y)} = \langle 0 | \, T u_a(x) \, \overline{u}_b(y) \, | 0 \rangle, \tag{4.135}$$

in momentum space given by

$$D_{ab}(Q) = \frac{\delta_{ab}}{Q^2 + i\varepsilon} \tag{4.136}$$

and visualized by the graphical symbol of a dotted line

$$Q$$
$$i \, D_{ab}(Q) \qquad a \; \cdots > \cdots \; b$$

with an arrow pointing into the direction of the momentum Q.

Furthermore, $\mathcal{L}_{\text{ghost}}$ contains trilinear interaction terms of each two scalar fields and one gauge field. By these couplings the ghost propagators occur in all closed loops with W propagators and trilinear self couplings. The structure of the ghost–gauge field interactions takes care that the ghosts cancel the unphysical components of the W fields.

For this cancellation a peculiarity of the ghost fields is required: though scalar fields, they develop an extra minus sign in closed loops as fermions do, where the sign originates from the anti-commutativity of the fermion fields. Formally the unorthodox sign of ghost loops comes about when the ghost fields are defined as anti-commuting variables (Grassman variables). This is not a violation of the spin–statistics theorem since u_a and $\overline{u}_a$ are unphysical degrees of freedom occuring only internally in loops but never as external fields or particles.

4.4.4.1 *BRST symmetry*

For the total Lagrangian, with $\mathcal{L}$ from Eq. (4.115),

$$\mathcal{L} + \mathcal{L}_{\text{fix}} + \mathcal{L}_{\text{ghost}}, \tag{4.137}$$

it is still true that the original symmetry under local gauge transformations applies only to the part $\mathcal{L}$ since $\mathcal{L}_{\text{fix}}$ is not invariant and gauge transformations for ghost fields are not defined. However, appropriate replacements and extensions of gauge transformations can be found that lead to a new symmetry of the total Lagrangian, the BRST symmetry, named after Becchi, Rouet, Stora, Tyutin [Becchi (1976); Tyutin (1975)].

In doing so, essentially the parameters $\delta\alpha_a(x)$ in infinitesimal local gauge transformations are identified with the ghost fields $u_a(x)$ according to[3]

$$\delta\alpha_a(x) = \theta\, u_a(x)$$

with an anticommuting constant θ (Lorentz scalar), and additional specific transformations are defined for the ghost fields. In case of gauge fixing like in Eq. (4.124) the BRST transformations for fermions, gauge fields and ghosts read as follows (with the notation $\phi \to \phi + \delta\phi$ for all fields),

$$\delta\psi_k = i\,\theta\, u_a \left(T_a\right)_{kl} \psi_l \tag{4.138}$$

$$\delta W_\mu^a = \frac{1}{g}\,\theta\left(\partial_\mu u_a + g\, f_{abc}\, W_\mu^b\, u_c\right) = \frac{1}{g}\,\theta\left(D_\mu^{\text{ad}}\right)_{ab} u_b$$

$$\delta\overline{u}_a = -\frac{1}{g\xi_a}\,\theta\, \partial^\mu W_\mu^a$$

$$\delta u_a = -\frac{1}{2}\,\theta\, f_{abc}\, u_b\, u_c.$$

Invariance under these transfomations replaces the original gauge symmetry. BRST invariance is the fundamental symmetry of the quantized theory and is crucial for the gauge independence of S-matrix elements for physical processes to all orders in perturbation theory.

---

[3] θ is an infinitesimal Grassman variable that anticommutes with the ghost fields u_a und $\overline{u}_a$.

4.4.5 *Perturbation theory and Feynman graphs*

The propagators of fermions, gauge bosons and ghosts follow from the quadratic part of the total Lagrangian (4.137). The residual part $\mathcal{L}_{\text{int}}$ describes the interaction by trilinear and quadrilinear coupling terms; it determines the various vertices for the interaction of gauge bosons with fermions and ghosts as well as gauge-boson self interactions. The vertices can be derived from the functional $i \int d^4x \, \mathcal{L}_{\text{int}}$, arising from the functional integral method as an access to quantization alternative to canonical quantization. An explanation would require a major excursion to formal quantum field theory and is thus beyond the scope of these introductory lectures.

For the perturbative expansion of the S-operator based on $\mathcal{H}_{\text{int}}$ this means that effectively one can set $\mathcal{H}_{\text{int}} = -\mathcal{L}_{\text{int}}$ even when derivatives are contained in $\mathcal{L}_{\text{int}}$, i.e. the extra terms in the Legendre transformation resulting from the derivatives can be ignored. The vertex corresponding to the coupling of n (generic) fields $\phi_1, \ldots \phi_n$ is then obtained from the expression

$$-i \int d^4x \, \langle 0 | \, \mathcal{H}_{\text{int}} \, | \, p_{\phi_1}, \ldots p_{\phi_n} \rangle = i \int d^4x \, \langle 0 | \, \mathcal{L}_{\text{int}} \, | \, p_{\phi_1}, \ldots p_{\phi_n} \rangle \qquad (4.139)$$

with formal momentum states $| p_{\phi_j} \rangle$ referring to the fields ϕ_j, after separating off the wave functions.

The more formal procedure using the vertex functional shall be illuminated in terms of a neutral scalar field $\phi(x)$ with self interactions,

$$\mathcal{L}_{\text{int}} = -\kappa \, \phi^3 - \lambda \, \phi^4, \qquad (4.140)$$

which describes the self-couplings of the Higgs field in the frame of the electroweak Standard Model (see Sec. 6.6). The functional for the vertices is given by

$$i \, \Gamma[\phi] = i \int d^4x \, \mathcal{L}_{\text{int}} = -i \int d^4x \, \left(\kappa \, \phi(x)^3 + \lambda \, \phi(x)^4 \right). \qquad (4.141)$$

The 3-point vertex is obtained as the functional derivative

$$i \, \Gamma^{(3)}(x_1, x_2, x_3) = i \, \left. \frac{\delta^3 \Gamma[\phi]}{\delta\phi(x_1) \, \delta\phi(x_2) \, \delta\phi(x_3)} \right|_{\phi=0}$$

and accordingly the 4-point vertex via

$$i \, \Gamma^{(4)}(x_1, x_2, x_3, x_4) = i \, \left. \frac{\delta^4 \Gamma[\phi]}{\delta\phi(x_1) \, \delta\phi(x_2) \, \delta\phi(x_3) \, \delta\phi(x_4)} \right|_{\phi=0}.$$

Perfoming the computation yields

$$i\,\Gamma^{(3)}(x_1, x_2, x_3) \qquad = -\,i\,3!\,\kappa\,\delta^4(x_1 - x_3)\,\delta^4(x_2 - x_3),$$

$$i\,\Gamma^{(4)}(x_1, x_2, x_3, x_4) = -\,i\,4!\,\lambda\,\delta^4(x_1 - x_4)\,\delta^4(x_2 - x_4)\,\delta^4(x_3 - x_4).$$

By Fourier transformation into momentum space

$$\tilde{\Gamma}^{(n)}(p_1, \ldots p_n) = \int d^4x_1 \ldots \int d^4x_n\; e^{ip_1 x_1} \ldots e^{ip_n x_n}\,\Gamma^{(n)}(x_1, \ldots x_n)$$

$$= (2\pi)^4\,\delta^4(p_1 + \cdots + p_n)\,\Gamma^{(n)}(p_1, \ldots p_n)$$

one obtains for $n = 3, 4$ the vertices

$$i\,\Gamma^{(3)}(p_1, p_2, p_3) \qquad = -\,i\,3!\,\kappa \tag{4.142}$$

$$i\,\Gamma^{(4)}(p_1, p_2, p_3, p_4) = -\,i\,4!\,\lambda$$

and the corresponding Feynman rules for the scalar self couplings

$$-\,i\,3!\,\kappa$$

$$-\,i\,4!\,\lambda$$

together with momentum conservation at each of the vertices. By convention, scalar fields are graphically displayed by dashed lines.

Chapter 5

Quantum Chromodynamics

Quantum chromodynamics (QCD) is the fundamental theory of the strong interaction, at the level of the building blocks of hadronic matter, the quarks. It is a gauge invariant quantum field theory with the gauge group $SU(3)$.

5.1 Formulation of QCD

Quarks occur in 6 specific versions, $q_f = u, d, s, c, b, t$, which differ with respect to mass, electromagnetic charge, and isospin. These properties are summarized under the name *flavour*. On top of the flavour degrees of freedom, an additional degree of freedom, named *colour*, exists that assigns a specific colour charge to each quark. While the flavour properties group the quarks into doublets and are relevant for the electromagnetic and weak interactions, the colour degree of freedom is responsibe for the strong interaction and makes each of the q_f to a triplet with different coulour charges, as displayed in Fig. 5.1. The masses of the q_f are of electroweak origin (see Sec. 6.4), for the strong interaction they are just given flavour parameters and hence they are equal for all the components of a colour triplet.

To each quark species for fixed flavour a colour triplet of spinor fields is assigned,

$$\psi = \begin{pmatrix} \psi_1 \\ \psi_2 \\ \psi_3 \end{pmatrix}, \quad \overline{\psi} = (\overline{\psi}_1, \overline{\psi}_2, \overline{\psi}_3) \tag{5.1}$$

where ψ_k are 4-component Dirac spinors and $\overline{\psi}_k$ the corresponding adjoint spinors. Since all ψ_k describe quarks of the same mass m, the free

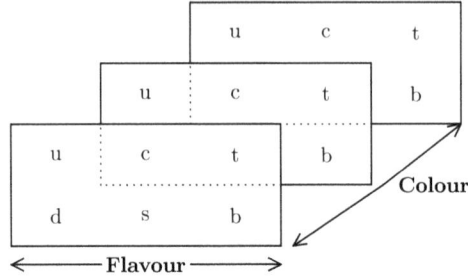

Fig. 5.1   Flavour and colour degrees of freedom of quarks.

Lagrangian can be written as follows,

$$\mathcal{L}_0 = \overline{\psi}\big(i\gamma^\mu\partial_\mu - m\big)\psi. \tag{5.2}$$

$\mathcal{L}_0$ is invariant under global gauge transformations of the group $SU(3) = SU(3)_C$, the gauge group of the strong interaction,

$$\psi' = e^{i\alpha_a T_a}\,\psi, \quad T_a = \frac{1}{2}\lambda_a \quad (a = 1,\ldots,8) \tag{5.3}$$

with generators T_a in the fundamental representation of $SU(3)$, given by Eq. (4.90). They fulfill the Lie algebra (4.85) with the structure constants f_{abc} listed in Tab. 4.1.

The following steps can be taken over from Sec. 4.4 almost literally. Invariance under local gauge transformations is achieved by introducing a covariant derivative with 8 gauge fields, denoted as *gluon fields*, and gluon dynamics is determined by means of the field strength tensor squared.

- Covariant derivative

$$D_\mu = \partial_\mu - i\,g_s T_a\,G_\mu^a \tag{5.4}$$

with the notations

G_μ^a :   gauge fields = gluon fields $(a = 1,\ldots,8)$
g_s :   coupling constant of the strong interaction

- Field strengths

$$G_{\mu\nu}^a = \partial_\mu G_\nu^a - \partial_\nu G_\mu^a + g_s f_{abc}\,G_\mu^b\,G_\nu^c \tag{5.5}$$

- Gauge invariant Lagrangian

$$\mathcal{L} = \overline{\psi}\big(i\gamma^\mu D_\mu - m\big)\psi - \frac{1}{4}\,G_{\mu\nu}^a\,G^{a,\mu\nu} \tag{5.6}$$

- Gauge-fixing term

$$\mathcal{L}_{\text{fix}} = \sum_a \frac{1}{2\xi} \left(\partial^\mu G_\mu^a \right)^2 \tag{5.7}$$

- Ghost Lagrangian

$$\mathcal{L}_{\text{ghost}} = \left(\partial^\mu \overline{u}_a \right) \left(\partial_\mu u_a \right) + g_s f_{abc} \left(\partial^\mu \overline{u}_a \right) G_\mu^b \, u_c \tag{5.8}$$

with a pair of ghost fields $u_a, \overline{u}_a$ for each gluon field G_μ^a.

In the physical reality one has to sum over the various flavours, with colour triplets ψ_f and masses m_f for $f = u, d, \ldots, t$. In this way the Lagrangian of QCD is accomplished,

$$\boxed{\mathcal{L}_{\text{QCD}} = \sum_f \overline{\psi}_f \left(i\gamma^\mu D_\mu - m_f \right) \psi_f - \frac{1}{4} G_{\mu\nu}^a \, G^{a,\mu\nu} + \mathcal{L}_{\text{fix}} + \mathcal{L}_{\text{ghost}}.} \tag{5.9}$$

Instead of the strong coupling constant g_s mostly the quantity

$$\alpha_s = \frac{g_s^2}{4\pi} \tag{5.10}$$

is used, the *fine structure constant of the strong interaction*, following the terminology of QED. It is the only free parameter of QCD since quark masses have to be considered parameters of the weak interaction. The value of α_s has to be determined experimentally by measurements of suitable observables. We will postpone the discussion of the strong coupling constant and refer to the later Sec. 5.3.

The interactions contained in $\mathcal{L}_{\text{QCD}}$, the quark–gluon interaction, the gluon self couplings, and the gluon–ghost interaction, can be translated into a list of vertices which together with the propagators and wave functions form the Feynman rules of QCD.

With these rules for the building blocks of Feynman graphs, the S-matrix elements for processes with quarks and gluons can be constructed. They form the basis for the calculation of cross sections of high-energetic scattering reactions directed by the dynamics of the strong interaction. This approach is known as *perturbative QCD*; it will be presented in the next section and illustrated by means of various examples.

5.1.1 *Feynman rules*

Fermions are represented graphically by lines with an arrow, like in QED, however, they carry an additional colour index. External wave functions of quarks and antiquarks are given by the spinors u, v for all colour indices,

External gluon lines refer, for all colour indices, to transverse polarization vectors ϵ_μ corresponding to helicities ± 1, like in the case of photons.

The gluon propagators as gauge-field propagators are given in Feynman gauge by the expression in Eq. (4.123), and the respective ghost propagators can be found in Eq. (4.136). The propagators of quarks are free fermion propagators and thus are of the same type as in QED, carrying merely an additional colour index.

New elements in the Feynman rules are the vertices for the non-Abelian self-interaction terms of the gluons and for the gluon–ghost interaction. Besides the coupling constant g_s, the structure constants f_{abc} enter the vertices; they are listed in Table 4.1. Another new element appears in the gluon–quark vertex, namely the matrix of the colour generators T_a in the fundamental representation, due to the structure of the quark currents which has been specified in Eq. (4.102).

The various interaction terms are represented by the following list of vertices (with indices $i, j = 1, 2, 3$ and $a, b, c, d = 1, \ldots, 8$).

$$i\, g_s \, (T_a)_{ij}\, \gamma^\mu$$

$$g_s\, f_{abc}\left[g^{\mu\nu}\,(k-p)^\rho + g^{\nu\rho}\,(p-q)^\mu + g^{\rho\mu}\,(q-k)^\nu \right]$$

$$-i\, g_s^2\left[f_{abc} f_{cde}\,(g^{\mu\rho} g^{\nu\sigma} - g^{\mu\sigma} g^{\nu\rho}) \right.$$
$$+ f_{ace} f_{bde}\,(g^{\mu\nu} g^{\rho\sigma} - g^{\mu\sigma} g^{\rho\nu})$$
$$\left. + f_{ade} f_{bce}\,(g^{\mu\nu} g^{\rho\sigma} - g^{\mu\rho} g^{\sigma\nu}) \right]$$

$$-\,g_s\, f_{abc}\, p^\mu$$

5.2   QCD Processes at High Energies

The calculation of S-matrix elements and cross sections within the framework of perturbative QCD is performed with the help of the Feynman rules for quarks and gluons. The applicability of perturbation theory requires that the strong coupling constant g_s, respectively the fine structure constant α_s, is sufficiently small, which for the strong interaction at a first glance sounds contradictory. However, as the discussion in the subsequent Sec. 5.3 will show, the requirement of a small coupling constant is fulfilled for processes at high energies and with large momentum transfer. Hence, high-energy reactions can be described consistently by the application of perturbative QCD.

The method of Feynman diagrams treats quarks and gluons as free particles, while in physics reality they occur exclusively as constituents of hadrons. The connection between the two aspects is established by the following rules.

- Quarks and gluons in the initial state are available as *partons* in nucleons and act like free particles in processes with large momentum transfer (see Sec. 5.4).
- Quarks and gluons in the final state turn into bunches of hadrons that are labeled as *jets*. For example, an outgoing $q\bar{q}$ pair results in two jets,

The jet axis obtained from the sum of the momenta of the hadrons inside a jet corresponds to the momentum direction of the primarily produced quark or gluon.

The following examples will illustrate the procedure for the calculation of amplitudes and cross sections.

(i) Electron–positron annihilation into quark–antiquark–gluon

Three-jet production $e^- + e^+ \to q + \bar{q} + G$ corresponds to the diagrams

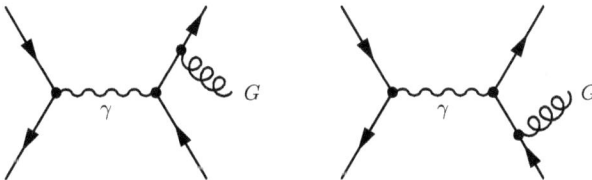

A basic difference with respect to QED consists in the appearance of colour factors. With the colour assignement q_k, $\bar{q}_l$ for the quarks and antiquarks and G_a for the gluons, the matrix element following from the Feynman rules has the structure

$$\mathcal{M} = g_s (T_a)_{kl} M \tag{5.11}$$

where M denotes the part without colour factors. Accordingly one obtains for the squared matrix element

$$|\mathcal{M}|^2 = g_s^2 (T_a)_{kl} (T_a)^*_{kl} |M|^2. \tag{5.12}$$

Since the colour degrees of freedom in the final state are not observed (quarks and gluons turn into colour-neutral hadrons) one has to sum over the indices a, k, l,

$$\sum_{a,k,l} |\mathcal{M}|^2 = g_s^2 \sum_{a=1}^{8} \sum_{k,l=1}^{3} (T_a)_{kl} (T_a)_{lk} |M|^2$$

$$= g_s^2 \sum_{a=1}^{8} \sum_{k=1}^{3} (T_a T_a)_{kk} |M|^2 = g_s^2 \sum_{a=1}^{8} \mathrm{Tr}(T_a T_a) |M|^2$$

$$= g_s^2 \sum_{a=1}^{8} \frac{1}{2} |M|^2 = g_s^2 \cdot 4 |M|^2. \tag{5.13}$$

In the next step one has to sum over the final-state helicities and average over the initial-state helicities, as described in Sec. 3.7 (the polarization sum (3.89) for photons given there is also applicable for gluons). The resulting spin- and colour-averaged squared matrix element

$$\overline{|\mathcal{M}|^2} = 4 g_s^2 \, \overline{|M|^2}$$

yields the differential 3-particle cross section according to Eq. (3.72),

$$d\sigma = \frac{1}{64 \pi^5 s} \overline{|\mathcal{M}|^2} \frac{d^3 p_1}{2 p_1^0} \frac{d^3 p_2}{2 p_2^0} \frac{d^3 p_3}{2 p_3^0} \delta^4 (P_i - p_1 - p_2 - p_3) \tag{5.14}$$

for the further evaluation in the CMS. p_1 and p_2 denote the momenta of q and $\bar{q}$, $s = (2E)^2$, and $P_i^0 = 2E$, $\vec{P}_i = 0$. E is the energy of the incoming electron and positron in the CMS. Treating all particles as massless and integrating over the angles yields the twofold differential cross section with respect to $E_1 = p_1^0$ and $E_2 = p_2^0$ as follows,

$$\frac{d^2\sigma}{dx_1\, dx_2} = \frac{2\alpha_s}{3\pi} \frac{x_1^2 + x_2^2}{(1 - x_1)(1 - x_2)}\, \sigma_0 \qquad (5.15)$$

with $x_1 = E_1/E$, $x_2 = E_2/E$. The quantity σ_0 is the cross section for the production of hadrons via $e^+e^- \to q\bar{q}$ given in (3.94). The distribution (5.15) provides a possibility for the experimental determination of the coupling α_s.

(ii) Top-quark pair production in quark–antiquark annihilation

The matrix element of the process $q + \bar{q} \to t + \bar{t}$ corresponds to a single Feynman diagram,

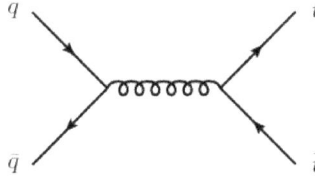

with the colour assignment $t_k, \bar{t}_l, q_i, \bar{q}_j$ for the external fermions and a for the internal gluon line, analytically given by

$$\mathcal{M} = g_s^2 \sum_a (T_a)_{kl} (T_a)_{ji}\, M, \qquad M = \frac{i}{s} (\bar{v}\gamma^\mu u)(\bar{u}'\gamma_\mu v'). \qquad (5.16)$$

The colour-free part M coincides up the factor e^2 with the matrix element for $e^+e^- \to \mu^+\mu^-$ in (3.57). Squaring yields the expression

$$|\mathcal{M}|^2 = g_s^4 \sum_{a,b} (T_a)_{kl} (T_a)_{ji} (T_b)_{kl}^* (T_b)_{ji}^* |M|^2$$

$$= g_s^4 \sum_{a,b} (T_a)_{kl} (T_b)_{lk} (T_a)_{ji} (T_b)_{ij} |M|^2. \qquad (5.17)$$

When colour summation is done one has to take into account that the colour degrees of freedom in the intitial state are averaged, i.e. for quarks and antiquarks a factor $1/3$ has to be attached in either case,

$$|\mathcal{M}|^2 \to \frac{1}{3} \cdot \frac{1}{3} \sum_{i,j=1}^{3} \sum_{k,l=1}^{3} |\mathcal{M}|^2. \qquad (5.18)$$

Yet the spin summation for $|M|^2$ has to be performed, then the colour- and spin-averaged matrix element squared is obtained,

$$\overline{|\mathcal{M}|^2} = g_s^4 \frac{1}{9} \sum_{i,j} \sum_{k,l} \sum_{a,b} (T_a)_{kl} (T_b)_{lk} (T_a)_{ji} (T_b)_{ij} \overline{|M|^2}$$

$$= g_s^4 \frac{1}{9} \sum_{a,b} \text{Tr}(T_a T_b) \text{Tr}(T_a T_b) \overline{|M|^2}$$

$$= g_s^4 \frac{1}{9} \sum_{a,b} \frac{1}{2}\delta_{ab} \cdot \frac{1}{2}\delta_{ab} = g_s^4 \frac{1}{9} \cdot \frac{1}{4} \cdot 8 \,\overline{|M|^2} = \frac{2}{9} g_s^4 \,\overline{|M|^2}. \qquad (5.19)$$

The calculation of the cross section is done as for $e^+ e^- \to \mu^+ \mu^-$ in Sec. 3.7, however, the finite mass m_t of the top quark has to be kept. According to Eq. (3.75) one obtains the differential cross section in the CMS, with $s = (p_q + p_{\bar{q}})^2$,

$$\frac{d\sigma}{d\Omega} = \frac{1}{64\pi^2 s} \sqrt{1 - \frac{4m_t^2}{s}} \cdot \overline{|\mathcal{M}|^2}$$

$$= \frac{1}{64\pi^2 s} \sqrt{1 - \frac{4m_t^2}{s}} \cdot \frac{2}{9} g_s^4 \left(1 + \cos^2\theta + \frac{4m_t^2}{s} \sin^2\theta\right) \qquad (5.20)$$

and after integration over $d\Omega$ also the integrated cross section,

$$\sigma_{q\bar{q} \to t\bar{t}}(s) = \frac{4\pi\alpha_s^2}{3s} \sqrt{1 - \frac{4m_t^2}{s}} \cdot \frac{2}{9} \left(1 + \frac{2m_t^2}{s}\right). \qquad (5.21)$$

(iii) Top-quark pair production in gluon–gluon fusion

The gluon-fusion process $G + G \to t + \bar{t}$ has a more complicated matrix element, displayed by several Feynman graphs

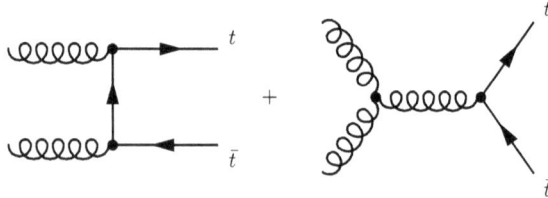

where a further diagram resulting from the left contribution by exchange of t and $\bar{t}$ is not shown explicitly. According to the momentum-squared t, u, s

at the respective internal line, the matrix element is decomposed into three contributions,

$$\mathcal{M} = \mathcal{M}_t + \mathcal{M}_u + \mathcal{M}_s. \tag{5.22}$$

With the colour assignment $t_k, \bar{t}_l$ for the external fermions, a, b for the external gluons, c for the internal gluon line, and m for the internal fermion line, one finds the colour structure of each part,

$$\mathcal{M}_t = g_s^2 \sum_m (T_a)_{km} (T_b)_{ml} \cdot M_t,$$

$$\mathcal{M}_u = g_s^2 \sum_m (T_b)_{km} (T_a)_{ml} \cdot M_u, \tag{5.23}$$

$$\mathcal{M}_s = g_s^2 \sum_c f_{abc} (T_c)_{kl} \cdot M_s.$$

M_t, M_u, M_s denote once again the colour-free parts of the contributions. Performing colour summation one has to take into accout that one has to average over the gluons in the intital state, i.e. a factor $1/8$ for each gluon has to be attached,

$$|\mathcal{M}|^2 \rightarrow \frac{1}{8 \cdot 8} \sum_{a,b=1}^{8} \sum_{k,l=1}^{3} |\mathcal{M}|^2. \tag{5.24}$$

Concerning spin summation, gluons are treated like photons, using the polarization sum (3.89) and assigning a factor $1/2$. With the notation for the spin-averaged quadratic terms $(r, r' = s, t, u)$

$$M_{rr} = \overline{|M_r|^2}, \quad M_{rr'} = 2\overline{\mathrm{Re}(M_r M_{r'}^*)}, \tag{5.25}$$

the spin- and colour-averaged matrix element squared can be written as follows,

$$\overline{|\mathcal{M}|^2} = \frac{g_s^4}{64} \left(\left\langle \frac{16}{3} \right\rangle M_{tt} + \left\langle \frac{16}{3} \right\rangle M_{uu} + \langle 12 \rangle M_{ss} \right.$$
$$\left. + \left\langle -\frac{2}{3} \right\rangle M_{tu} + \langle 6 \rangle M_{ts} + \langle 6 \rangle M_{us} \right)$$

where the brackets $\langle \ \rangle$ indicate that the coefficients originate from colour summation. Then the differential cross section in the CMS is obtained according to

$$\frac{d\sigma}{d\Omega} = \frac{1}{64\pi^2 s} \sqrt{1 - \frac{4m_t^2}{s}} \, \overline{|\mathcal{M}|^2} \, .$$

It is convenient to use besides the CMS-energy $\sqrt{s}$ also the other invariant Mandelstam variables t and u according to Eq. (3.90), with $s + t + u = 2m_t^2$,

$$t = m_t^2 - \frac{s}{2}\left(1 - \beta\cos\theta\right), \quad u = m_t^2 - \frac{s}{2}\left(1 + \beta\cos\theta\right), \quad \beta = \sqrt{1 - \frac{4m_t^2}{s}},$$

and to replace $\cos\theta$ by t in the cross section, yielding

$$\begin{aligned}
\frac{d\sigma}{dt} &= \frac{1}{16\pi s^2} \overline{|\mathcal{M}|^2} \\
&= \frac{\pi\alpha_s^2}{64 s^2}\left(\left\langle\frac{16}{3}\right\rangle M_{tt} + \left\langle\frac{16}{3}\right\rangle M_{uu} + \langle 12\rangle M_{ss}\right. \\
&\quad \left. + \left\langle-\frac{2}{3}\right\rangle M_{tu} + \langle 6\rangle M_{ts} + \langle 6\rangle M_{us}\right).
\end{aligned} \tag{5.26}$$

The calculation of the various terms is somewhat lengthy but straightforward and can be performed applying the tools presented in Sec. 3.7. Without entering into details the result is listed here,

$$M_{tt} = 2\,\frac{(t - m_t^2)(u - m_t^2) - 2m_t^2(t - m_t^2) - 4m_t^4}{(t - m_t)^2},$$

$$M_{uu} = 2\,\frac{(t - m_t^2)(u - m_t^2) - 2m_t^2(u - m_t^2) - 4m_t^4}{(u - m_t)^2},$$

$$M_{ss} = 4\,\frac{(t - m_t^2)(u - m_t^2)}{s^2},$$

$$M_{ts} = 4\,\frac{m_t^4 - t(s + t)}{s(t - m_t^2)},$$

$$M_{us} = 4\,\frac{m_t^4 - u(s + u)}{s(u - m_t^2)},$$

$$M_{tu} = 4\,\frac{m_t^2(s - 4m_t^2)}{(t - m_t^2)(u - m_t^2)}.$$

$$\tag{5.27}$$

With these entries the integrated cross section follows from Eq. (5.26),

$$\sigma_{GG \to t\bar{t}}(s) = \frac{\pi \alpha_s^2}{64s} \left[\langle 12 \rangle \frac{2 + \rho}{3} \beta + \left\langle \frac{16}{3} \right\rangle \left((4 + 2\rho) \ln \frac{1 + \beta}{1 - \beta} - 4(1 + \rho)\beta \right) \right.$$

$$\left. + \langle 6 \rangle \left(2\rho \ln \frac{1 + \beta}{1 - \beta} - 4(1 + \rho)\beta \right) + \left\langle -\frac{2}{3} \right\rangle 2\rho (1 - \rho) \ln \frac{1 + \beta}{1 - \beta} \right]$$

$$(5.28)$$

where $\rho = 4m_t^2 / s$.

5.2.1 *Processes at hadron colliders*

The quantitites $\sigma_{q\bar{q} \to t\bar{t}}$ and $\sigma_{GG \to t\bar{t}}$ in *(ii)* and *(iii)* represent the cross sections for the elementary partonic processes of $t\bar{t}$ production at hadron colliders, the proton–antiproton collider ($P\bar{P}$) TEVATRON (from 1983 until 2011 at the Fermi National Laboratory) and the proton–proton collider (PP) LHC (since 2010 at CERN). In order to get to the measureable cross section for $t\bar{t}$ production, the partonic cross sections have to be weighted by the corresponding parton distributions and integrated over the parton momenta. Parton distributions are defined as follows:

- $f_q^h(x)$, $f_{\bar{q}}^h(x)$ is the probability density for a quark or an antiquark inside the hadron $h = P, \bar{P}$ carrying the momentum fraction x of the hadron momentum.
- $f_G(x)$ is the probabilty density for a gluon inside the hadron P or $\bar{P}$ carrying the momentum fraction x of the hadron momentum.

For collisions of two hadrons h_1 und h_2 with momenta P_1 und P_2 and total energy $\sqrt{S}$ in the CMS, with $S = (P_1 + P_2)^2$, the observable hadronic cross sections are calculated in two steps.

(1) Calculation of the cross sections $\sigma_{ij \to X}(s)$ for the partonic reactions $ij \to X$ with partons $i, j = q, \bar{q}, G$ and energy $\sqrt{s}$ in the parton CMS, with $s = (p_i + p_j)^2$.
(2) Calculation of the hadronic cross section by multiplication with the corresponding parton distributions and subsequent integration over the parton momenta $p_i = x_1 P_1$, $p_j = x_2 P_2$ and summation over all

contributing partons,

$$\sigma_{h_1 h_2 \to X}(S) = \sum_{i,j} \int_0^1 dx_1 \int_0^1 dx_2 \, f_i^{h_1}(x_1) \, f_j^{h_2}(x_2) \, \sigma_{ij \to X}(x_1 x_2 S).$$

$$(5.29)$$

Since all particles are treated as massless, the following relations hold,

$$s = (p_i + p_j)^2 = 2(p_i p_j) = x_1 x_2 \cdot 2(P_1 P_2) = x_1 x_2 S.$$

For the concrete example of top-quark pair production at the LHC the hadronic cross section follows from the contributions of GG fusion and $q\bar{q}$ annihilation with $q = u, d, s, c, b$,

$$\sigma_{PP \to t\bar{t}}(S) = \int_0^1 dx_1 \int_0^1 dx_2 \, f_G(x_1) \, f_G(x_2) \, \sigma_{GG \to t\bar{t}}(x_1 x_2 S)$$

$$+ \sum_q \int_0^1 dx_1 \int_0^1 dx_2 \left[f_q^P(x_1) \, f_{\bar{q}}^P(x_2) + f_{\bar{q}}^P(x_1) \, f_q^P(x_2) \right]$$

$$\times \sigma_{q\bar{q} \to t\bar{t}}(x_1 x_2 S)$$

with the partonic cross sections from Eqs. (5.21) and (5.28). At the LHC with the high energy $\sqrt{S} = 13\,\mathrm{TeV}$ the gluon-fusion subprocess provides the main contribution to the hadronis cross section. At the $P\bar{P}$ collider TEVATRON with $\sqrt{S} = 2\,\mathrm{TeV}$, on the other hand, the $q\bar{q}$ subprocesses dominate the hadronic cross section,

$$\sigma_{P\bar{P} \to t\bar{t}}(S) = \sum_q \int_0^1 dx_1 \int_0^1 dx_2 \left[f_q^P(x_1) \, f_{\bar{q}}^{\bar{P}}(x_2) + f_{\bar{q}}^P(x_1) \, f_q^{\bar{P}}(x_2) \right]$$

$$\times \sigma_{q\bar{q} \to t\bar{t}}(x_1 x_2 S).$$

An explanation can be found at the end of Sec. 5.4.

Parton distributions are essentially experimentally determined, primarily extracted from deep-inelastic lepton–nucleon scattering. They are documented in terms of convenient parametrizations and are also available as computer programs for concrete calculations (see e.g. [Zyla (2020)] for more information).

5.3 Running Coupling Constant

The applicability of perturbation theory to QCD processes at high energies requires a sufficiently weak coupling to make sure that the contributions of increasing order become smaller and smaller. In this section a reason will be given why this requirement is fulfilled, by studying the behaviour of the strong coupling constant in terms of α_s with respect to the dependence on the energy scale. The feature that the value of a coupling constant does depend at all on the energy of the considered processes is a consequence of higher-order terms with loop diagrams. Only at lowest order, like in the examples of the previous section, the coupling constant is actually a constant.

5.3.1 *Fine structure constant of QED*

At first the simpler case of QED is disussed where the coupling constant is given by the elementary charge e or the fine structure constant α, respectively. According to Sec. 3.5 the interaction between two electrons in lowest order is described by the exchange of a photon and the matrix element is represented by a Feynman diagram without any loop, carrying the momentum transfer q in the photon propagator. At the next order loop diagrams occur, in particular the vacuum polarization as an insertion in the photon propagator which can be resummed iteratively,

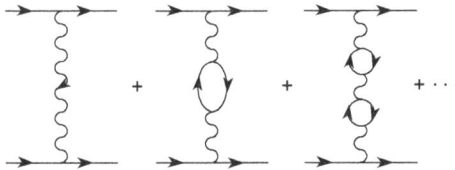

With the notation Π for the vacuum polarization, insertion of Π modifies the matrix element of lowest order according to the replacement

$$\frac{e^2}{q^2} \rightarrow \frac{e^2}{q^2} + \frac{e^2}{q^2} \cdot (-\Pi) \tag{5.30}$$

and the matrix element with iterated insertions yields a geometrical progression,

$$\mathcal{M} \sim \frac{e^2}{q^2} \left(1 - \Pi + \Pi^2 - \cdots\right) = \frac{1}{q^2} \frac{e^2}{1 + \Pi}. \tag{5.31}$$

The vacuum polarization is calculated as the sum of the fermion–antifermion loops of charged leptons and quarks, with charges Q_f and masses m_f, represented by the diagrams with internal e, μ, τ, q fermion lines

The calculation is done applying the rules given in Secs. 3.7 and 3.8 with integration over the loop momentum. The result has a logarithmic dependence on the propagator momentum squared,

$$\Pi(Q^2) = -\frac{\alpha}{3\pi} \sum_{f=e,\mu,\dots} Q_f^2 \left(\log \frac{Q^2}{m_f^2} + \cdots \right) \tag{5.32}$$

for $Q^2 \equiv |q^2| \gg m_f^2$. The terms indicated by $\cdots$ are momentum-independent and not needed for the considerations of this section. In the resummed version (5.31) of the matrix element one can interprete the quantity

$$\frac{e^2}{1 + \Pi(Q^2)}$$

as the square of an effective charge, corresponding to an effective fine structure constant

$$\alpha(Q^2) = \frac{\alpha}{1 + \Pi(Q^2)} \tag{5.33}$$

at the scale Q^2 (running coupling constant). With the choice $\alpha(Q_0^2)$ at a fixed Q_0^2 as an initial value one obtains the expression

$$\alpha(Q^2) = \frac{\alpha(Q_0^2)}{1 + \frac{\alpha(Q_0^2)}{4\pi} \beta_0 \log \frac{Q^2}{Q_0^2}} \tag{5.34}$$

with the coefficient

$$\beta_0 = -\frac{4}{3} \sum_f Q_f^2. \tag{5.35}$$

Because of the negative sign of β_0 the running coupling constant of QED increases for increasing Q^2.

The running coupling (5.34) is a solution of the differential equation

$$Q^2 \frac{\partial \alpha}{\partial Q^2} = -\frac{\beta_0}{4\pi} \alpha^2 \tag{5.36}$$

with the initial value $\alpha(Q_0^2)$. This is a special case of the evolution equation (also called renormalization group equation, RGE)

$$Q^2 \frac{\partial \alpha}{\partial Q^2} = \beta(\alpha) \tag{5.37}$$

for the scale dependence of the coupling which is determined by the β-function $\beta(\alpha)$. The β-function can be specified perturbatively as an expansion in powers of α,

$$\beta(\alpha) = -\frac{\alpha^2}{4\pi}\beta_0 - \frac{\alpha^3}{(4\pi)^2}\beta_1 - \frac{\alpha^4}{(4\pi)^3}\beta_2 - \cdots \qquad (5.38)$$

where the coefficients β_k have to be determined order by order in perturbation theory, i.e. β_0 is obtained from the one-loop diagrams for the vacuum polarization, β_1 from the two-loop diagrams, β_2 from the three-loop diagrams, etc. Hence, the one-loop result (5.34) is corrected accordingly by the multi-loop contributions. The solution of the RGE (5.37) with $\beta(\alpha)$ in n-th order corresponds to the resummation of the iterated insertions of the n-loop vacuum polarisation, formally as in (5.31), with $\Pi(Q^2)$ augmented by higher-order contributions. A justification of this systematic method is given by the renormalization group, a major topic of quantum field theory that is beyond the scope of these introductory lectures (see for example [Peskin (1995)]).

5.3.2 *Fine structure constant of QCD*

Now we turn to the strong analogon of the electron–electron interaction, the quark–quark interaction, which is described in the framework of perturbative QCD by single-gluon exchange at lowest order. Like in QED, the matrix element at the next order contains one-loop contributions to the gluon propagator

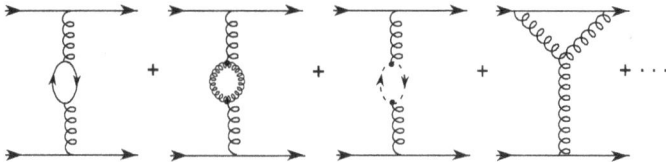

that form the vacuum polarization of the gluon. Also in analogy to the photon, fermionic contributions exist in terms of quark–antiquark loops; differently from photons, however, there are also bosonic contributions by gluon loops originating from the non-Abelian gauge self-couplings, supplemented by the ghost loops for the cancellation of the unphysical gluon polarizations, with couplings of the ghosts to gluons according to the Feynman rules in Sec. 5.1.

The fermionic contributions to the vacuum polarization are calculated, following the Feynman rules, as traces over the fermion loops and integration over the loop momentum, with an extra minus sign. Differences with respect to QED result from the replacement of the coupling $e \to g_s$ and charges $Q_f \to T_a$ as well as from the summation over the colour part of the quark loops in the gluon line (with colour index a),

$$e^2 Q_f^2 \to g_s^2 \sum_{k,l} (T_a)_{kl} (T_a)_{lk} = g_s^2 \operatorname{Tr} (T_a T_a) = \frac{g_s^2}{2}$$

for each quark-flavour degree of freedom $F = u, d, \dots$. Summation over the quark flavours has to be done on top, so that altogether the substitution

$$\sum_f e^2 Q_f^2 \to \sum_F \frac{g_s^2}{2} = g_s^2 \frac{n_F}{2}$$

has to be performed in Eq. (5.32) to obtain the fermionic part of the gluon vacuum polarisation. n_F denotes the number of different quark flavours ($n_F = 6$, but only when $Q^2 > 4m_t^2$, otherwise $n_F = 5$). The contribution of the quark loops to β_0 in QCD is obtained as the coefficient of $\log(Q^2)$ in the vacuum polarization with the appropriate normalization,

$$\Pi^{(\text{quarks})} = \frac{\alpha_s}{4\pi} \beta_0^{(q)} \left[\log(Q^2) + \cdots \right] \tag{5.39}$$

yielding

$$\beta_0^{(q)} = -\frac{2}{3} n_F, \tag{5.40}$$

with the same sign as in case of QED. A crucial difference with respect to QED, however, results from the non-Abelian structure of QCD, namely from the gluon and ghost loops. In combination they provide the gluonic contribution to the β-function

$$\beta_0^{(G)} = +11 \tag{5.41}$$

with the opposite sign of the fermionic contribution. In total, QCD predicts the coefficient

$$\beta_0^{\text{QCD}} = \beta_0^{(q)} + \beta_0^{(G)} = 11 - \frac{2}{3} n_F \tag{5.42}$$

with the property $\beta_0^{\text{QCD}} > 0$ for $33 - 2n_F > 0$, which is valid for $n_F \leq 6$ in any case. The running coupling constant of QCD thus is given by

$$\boxed{\alpha_s(Q^2) = \frac{\alpha_s(Q_0^2)}{1 + \frac{\alpha_s(Q_0^2)}{4\pi} \left(11 - \frac{2}{3} n_F\right) \log \frac{Q^2}{Q_0^2}}} \tag{5.43}$$

with $\alpha_s(Q_0^2)$ as input for a (in principle arbitrary) reference value Q_0^2.

In a non-Abelian theory the β-function for the gauge coupling is determined not only by the vacuum polarization of the gauge bosons but also by a specific part of the vertex correction, in case of the quark–quark interaction derived from the diagram with the 3-gluon vertex as the coefficient of a $\log(Q^2)$ term as well. Following (5.30) the modfication of g_s at the one-loop level is given by

$$g_s^2 \rightarrow g_s^2 \left[1 - \Pi^{\text{(quarks)}} - \Pi^{\text{(gluons)}} - \Pi^{\text{(ghosts)}} - V^{\text{(3gluon)}} \right]$$

with the contributions to the vacuum polarization from the quark loops in Eq. (5.39), from the gluon and ghost loops

$$\Pi^{\text{(gluons)}} = \frac{\alpha_s}{4\pi} \cdot \frac{19}{4} \cdot \left[\log(Q^2) + \cdots \right] \Rightarrow \beta_0^{\text{(gluons)}} = \frac{19}{4},$$

$$\Pi^{\text{(ghosts)}} = \frac{\alpha_s}{4\pi} \cdot \frac{1}{4} \cdot \left[\log(Q^2) + \cdots \right] \Rightarrow \beta_0^{\text{(ghosts)}} = \frac{1}{4},$$

and moreover from the vertex correction with the non-Abelian 3-gluon coupling

$$V^{\text{(3gluon)}} = \frac{\alpha_s}{4\pi} \cdot 6 \cdot \left[\log(Q^2) + \cdots \right] \Rightarrow \beta_0^{\text{(3gluon)}} = 6,$$

resulting in the total gluonic part of the β-function

$$\beta_0^{\text{(G)}} = \beta_0^{\text{(gluons)}} + \beta_0^{\text{(ghosts)}} + \beta_0^{\text{(3gluon)}} = 11$$

as stated above in Eq. (5.41). Resummation of the logarithms and fixing the initial value of α_s at Q_0^2 finally leads to the expression (5.43) for $\alpha_s(Q^2)$. The resummation corresponds to the iterated insertions of the vacuum polarization in the gluon propagator, in analogy to Eq. (5.31). That the vertex part is treated in the same way, on the other hand, is not immediately clear. The already mentioned systematic renormalization group method, however, identifies the vertex part $\beta_0^{\text{(3gluon)}}$ given above as the one-loop vertex contribution to the β-function of QCD. Hence, the RGE for the running coupling $\alpha_s(Q^2)$ at one-loop order is given by

$$Q^2 \frac{\partial \alpha_s}{\partial Q^2} = \beta(\alpha_s) = -\frac{\beta_0^{\text{QCD}}}{4\pi} \alpha_s^2 \tag{5.44}$$

with the solution (5.43). Besides $\beta_0 \equiv \beta_0^{\text{QCD}}$, the coefficients β_k in the expansion

$$\beta(\alpha_s) = -\frac{\alpha_s^2}{4\pi} \beta_0 - \frac{\alpha_s^3}{(4\pi)^2} \beta_1 - \frac{\alpha_s^4}{(4\pi)^3} \beta_2 - \cdots \tag{5.45}$$

are known up to β_4, hence the scale dependence of α_s is determined up to the order of 5 loops by means of Eq. (5.44). Though, an exact analytical solution exists only in terms of Eq. (5.43) at lowest order.

The positive sign of the β-function in QCD has far-reaching physical consequences. In the limit of large scales $Q^2 \to \infty$ one finds $\alpha_s(Q^2) \to 0$, which means that the coupling constant becomes small and thus the strong interaction becomes weak at high energies and large momentum transfer. This spectacular feature of the strong interaction is denoted as *asymptotic freedom*. It was discovered in 1973 by Gross, Wilczek, and Politzer [Gross (1973); Politzer (1973)] and recognized later by the Nobel prize in the year 2004. Asymptotic freedom justifies the applicability of perturbative QCD and gives the reason for the appearence of quasi-free partons in the nucleon for high energy scattering processes with large Q^2.

Experimentally the running of α_s has been confirmed by plenty of measurements performed by various experiments in different energy ranges up to the highest energies at the LHC, as displayed in Fig. 5.2. It has become customary to choose the scale of the Z-boson mass M_Z at 91 GeV as a reference value and to convert the values of α_s measured at other scales to this reference value. The world average resulting from this procedure is

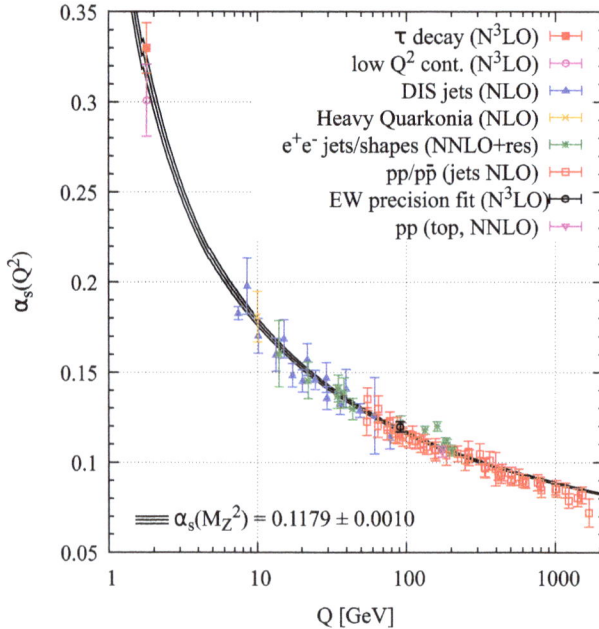

Fig. 5.2 Running coupling constant of the strong interaction and experimental data at different scales Q (from [Zyla (2020)]). The curves correspond to the theory prediction for the given value (with errors) of $\alpha_s(M_Z)$, based on the known β-function.

given by [Zyla (2020)]

$$\alpha_s(M_Z^2) = 0.1179 \pm 0.0010, \qquad (5.46)$$

it serves also as a numerical input for calculations of processes at much higher energies, like processes at the LHC. As an example we refer to the cross section of $t\bar{t}$ pair production calculated in Sec. 5.2: in the expressions (5.21) and (5.28) for the partonic cross sections of quark–antiquark annihilation and gluon–gluon fusion one has to specify α_s by $\alpha_s(Q^2)$ with $Q^2 = m_t^2$. Numerically the value of α_s is decreased by the running to 0.108.

In contrast to the high-energy regime with asymptotic freedom, α_s increases strongly for low Q^2 and diverges (formally) at a scale Λ where the denominator in Eq. (5.43) has a zero,

$$\Lambda^2 = Q_0^2 \exp\left[-\frac{12\pi}{(33 - 2n_F)\,\alpha_s(Q_0^2)} \right].$$

At one-loop order one obtains for $Q^2 = \Lambda^2$,

$$\frac{1}{\alpha_s(\Lambda^2)} - \frac{1}{\alpha_s(Q_0^2)} = -\frac{1}{\alpha_s(Q_0^2)} = \frac{33 - 2n_F}{12\pi} \log\frac{\Lambda^2}{Q_0^2},$$

and thus one can replace $\alpha_s(Q_0^2)$ by Λ according to

$$\frac{1}{\alpha_s(Q^2)} = \frac{1}{\alpha_s(Q_0^2)} + \frac{33 - 2n_F}{12\pi} \log\frac{Q^2}{Q_0^2} = \frac{33 - 2n_F}{12\pi} \log\frac{Q^2}{\Lambda^2},$$

yielding

$$\alpha_s(Q^2) = \frac{12\pi}{(33 - 2n_F) \log\frac{Q^2}{\Lambda^2}}. \qquad (5.47)$$

The divergence at Λ has to be understood in the sense that for Q^2 near Λ^2 perturbation theory is not applicable. For $Q^2 \approx \Lambda^2$ and $Q^2 < \Lambda^2$ alternative, non-perturbative, methods are required, such as the lattice approximation or effective field theories as low-energy limit of QCD, with typical applications for example in meson and baryon spectroscopy.

The numerical value of Λ depends on the number of flavours and, moreover, on the order of perturbation theory. For $n_F = 5$ and at 4-loop order one obtains, corresponding to $\alpha_s = 0.1181 \pm 0.0011$, the value

$$\Lambda = 210 \pm 14\,\text{MeV} \qquad (5.48)$$

(see [Tanabashi (2018)] for more information).

5.4 Parton Distributions

In Sec. 5.2 it was already pointed out that in high-energy scattering processes with large momentum transfer, so-called hard scattering processes, quarks and gluons appear as free particles, named partons, inside the nucleon and act as primary reaction partners in hadronic collisions, with characteristic momentum distributions that are universal, i.e. independent of the details of the scattering processes. These parton distributions have to be considered to be among the basic properties of nucleons.

5.4.1 *Parton model*

The concept of free partons was introduced by Feynman [Feynman (1969)], motivated by the results of experiments to deep-inelastic electron–nucleon scattering at SLAC [Breidenbach (1969)] that could be explained by scattering of electrons by pointlike spin-1/2 constituents of the nucleon, as displayed in Fig. 5.3.[1]

Historically the partons postulated by Feynman had nothing to do with the quarks introduced by Gell-Mann, which were mathematical objects with the aim to explain the hadron spectrum. Only in course of the subsequent decade, with further experiments to deep-inelastic lepton–nucleon scattering and to e^+e^- annihilation, when the colour degree of freedeom was

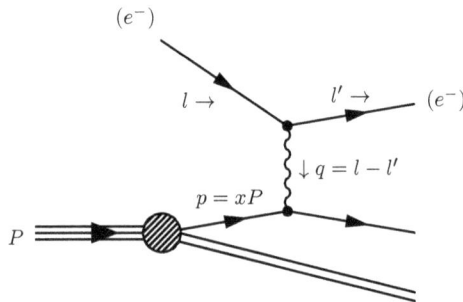

Fig. 5.3 Deep-inelastic electron–proton scattering in the parton model with photon exchange.

---

[1]Note that the presentation in Fig. 5.3 is not a Feynman graph; only the part describing the electron–quark scattering process corresponds to the Feynman diagram for the respective matrix element.

gradually established and QCD proved more and more successful, the spin-1/2 partons of the hard scattering processes could be identified with the quarks of the hadron spectrum.

According to the way how nucleons are composed of quarks, parton distributions refering to the various quark flavours are assigned to proton and neutron (antiproton follows from proton by the permutation $u \to \bar{u}, d \to \bar{d}$):

proton:	u, u, d	$+$	$u\bar{u}, d\bar{d}, s\bar{s}, c\bar{c}, b\bar{b}$
	valence quarks		sea quarks
neutron:	u, d, d	$+$	$u\bar{u}, d\bar{d}, s\bar{s}, c\bar{c}, b\bar{b}$
	valence quarks		sea quarks

The valence quarks represent the constituents of the respective nucleon providing the appropriate electromagnetic charge of proton and neutron. On top, there are quark–antiquark pairs of light flavours that contribute to the nucleon momentum as well; they are denoted as sea quarks. Gluons do not participate in deep-inelastic scattering processes directly, they carry, however, an important fraction of the nucleon momentum (about one half). At higher orders gluons take part in scattering reactions either, as will be explicated below in Sec. 5.4.3.

The parton distribution functions (the conventional abbreviation is PDF) and their notations were introduced in Sec. 5.2, for processes at hadron colliders. The following notations are common,

$$f_q^P(x) \equiv q(x), \quad f_{\bar{q}}^P(x) \equiv \bar{q}(x), \quad q = u, d, s, c, b, \tag{5.49}$$

and the following relations hold between proton P and neutron N,

$$f_d^N(x) = f_u^P(x) = u(x), \quad f_u^N(x) = f_d^P(x) = d(x) \tag{5.50}$$

as well as between proton and antiproton $\bar{P}$,

$$f_{\bar{u}}^{\bar{P}}(x) = f_u^P(x) = u(x), \quad f_{\bar{d}}^{\bar{P}}(x) = f_d^P(x) = d(x),$$
$$f_u^{\bar{P}}(x) = f_{\bar{u}}^P(x) = \bar{u}(x), \quad f_d^{\bar{P}}(x) = f_{\bar{d}}^P(x) = \bar{d}(x). \tag{5.51}$$

5.4.2 *Deep-inelastic scattering*

The most immediate access to the parton distributions is provided by deep-inelastic scattering of leptons by nucleons. As a specific example let us consider electron–proton scattering (see Fig. 5.3), with momenta l and l' for the incoming and outgoing lepton and momentum transfer $q^2 = (l - l')^2$. The notation *deep-inelastic* is for the kinematical region with $Q^2 \equiv -q^2 > (2\,\text{GeV})^2$, where the parton model applies and the primary processes proceed between the lepton and the quarks of the nucleon.

The matrix element for electron–quark scattering is graphically displayed by a Feynman diagram that is part of Fig. 5.3, with the quark fraction x of the proton momentum P, so that the incoming quark q carries the momentum $p = xP$. This is a QED process, and accordingly with the Feynman rules of Sec. 3.5 one obtains

$$\mathcal{M} = i\,\frac{e^2 Q_q}{t}\,\left[\bar{u}(l')\gamma^\mu u(l)\right]\left[\bar{u}(p')\gamma_\mu u(p)\right] \tag{5.52}$$

with the notations Q_q for the charge of the quark q and $t = (l - l')^2 = (p-p')^2$ for the momentum transfer mediated by the virtual photon. Setting all masses to zero and performing the usual steps of squaring the amplitude and averaging/summing over the initial-state and final-state helicities yields

$$\overline{|\mathcal{M}|^2} = \frac{e^4 Q_q^2}{t^2}\cdot 2(s^2 + u^2) \tag{5.53}$$

expressed in terms of t and the other two invariant variables

$$\begin{aligned}
s &= (l + p)^2 = (l' + p')^2 = 2lp = xS, \\
u &= (l - p')^2 = (l' - p)^2 = -2l'p = -x \cdot 2Pl'
\end{aligned} \tag{5.54}$$

where $S = (l + P)^2$ is the total energy squared in the electron–proton CMS. Accordingly, the variable s corresponds to the total energy squared in the electron–quark CMS. From Eq. (5.53) the differential cross section for eq scattering is obtained in an invariant way,

$$\frac{d\sigma^{eq}}{dt} = \frac{1}{16\pi s^2}\overline{|\mathcal{M}|^2} = \frac{2\pi\alpha^2}{t^2}\cdot\frac{s^2 + u^2}{s^2}\cdot Q_q^2.$$

Customarily the variable $Q^2 = -t$ (with $Q^2 > 0$) is used. The relation $s + t + u = 0$, valid for massless particles, implies $u = -s - t = -s + Q^2$ and thus

$$\frac{d\sigma^{eq}}{dQ^2} = \frac{2\pi\alpha^2}{Q^4}\left[1 + \left(1 - \frac{Q^2}{s}\right)^2\right]\cdot Q_q^2 \tag{5.55}$$

for the differential eq cross section. Multiplication by the probability $q(x)\,dx$ that the quark q in the proton carries the momentum fraction between x and $x + dx$, and summation over all contributing q yields

$$\sum_q \frac{d\sigma^{eq}}{dQ^2} q(x)\, dx = \frac{2\pi\alpha^2}{Q^4}\left[1 + \left(1 - \frac{Q^2}{s}\right)^2\right] \sum_q Q_q^2\, q(x)\, dx \equiv \frac{d^2\sigma^{eP}}{dx\, dQ^2}\, dx$$

and thus the 2-fold differential eP cross section is obtained as follows,

$$\boxed{\frac{d^2\sigma^{eP}}{dx\, dQ^2} = \frac{2\pi\alpha^2}{Q^4}\left[1 + \left(1 - \frac{Q^2}{s}\right)^2\right] \sum_q Q_q^2\, q(x)}. \qquad (5.56)$$

The quantitites x und Q^2 can be determined experimentally from the energy and the scattering angle of the scattered electron, in combination with S. For Q^2 this is obvious; concerning x one makes use of the kinematical relations (5.54), yielding

$$u = Q^2 - s = Q^2 - xS = -2xPl' \Rightarrow x = \frac{Q^2}{S - 2Pl'}. \qquad (5.57)$$

The deep-inelastic eP cross section can be expressed independently of the parton model by exploiting Lorentz invariance and current conservation, yielding a parametrization by means of two *structure functions* $F_1(x, Q^2)$ and $F_2(x, Q^2)$ that depend in general on the two variables x and Q^2,

$$\boxed{\frac{d^2\sigma^{eP}}{dx\, dQ^2} = \frac{2\pi\alpha^2}{Q^4}\left[2\left(\frac{Q^2}{s}\right)^2 F_1 + 2\left(1 - \frac{Q^2}{s}\right)\frac{1}{x}F_2\right]}. \qquad (5.58)$$

As a result of the experiments performed at SLAC for the first time, the structure functions were found to be independent of Q^2 and interconnected,

$$F_1 = F_1(x),\ F_2 = F_2(x) \quad \text{with } F_2(x) = 2xF_1(x). \qquad (5.59)$$

This behaviour, denoted as *scaling*, corresponds precisely to the situation that the electron is scattered by pointlike constituents of the proton with spin $1/2$. Inserting Eq. (5.59) into the parametrized cross section (5.58) reproduces the prediction (5.56) of the parton model with the identification

$$F_2(x) = x \sum_q Q_q^2\, q(x), \qquad (5.60)$$

providing the interpretation of the structure functions in terms of parton distributions. For the proton one obtains

$$\frac{1}{x} F_2^P = \frac{4}{9}\left[u(x) + \bar{u}(x)\right] + \frac{1}{9}\left[d(x) + \bar{d}(x)\right] + \frac{1}{9}\left[s(x) + \bar{s}(x)\right] + \cdots$$

For other targets like neutron or deuteron, analogous steps as for the proton yield the same form of the cross section as in Eq. (5.56), however, with different combinations of distribution functions,

$$\frac{1}{x} F_2^N = \frac{4}{9} \left[d(x) + \bar{d}(x)\right] + \frac{1}{9} \left[u(x) + \bar{u}(x)\right] + \frac{1}{9} \left[s(x) + \bar{s}(x)\right] + \cdots$$

$$\frac{1}{x} F_2^D = \frac{1}{2x} \left(F_2^P + F_2^N\right) = \frac{5}{18} \left[u(x) + \bar{u}(x) + d(x) + \bar{d}(x) + \cdots\right].$$

By measurements of the various structure functions independent information is achieved on the quark distributions and their x-dependence.

> Further structure functions are available from deep-inelastic neutrino scattering where the primary process of neutrino–quark scattering is mediated by the weak interaction, for example in $\nu_\mu P \to \mu^- X$ by exchange of a charged W boson (see Sec. 6.3.2).

In the discussion so far, electron–nucleon scattering is mediated by the electromagnetic interaction (at higher energies and Q^2 as well as for neutrino scattering also by the weak interaction), whereas the strong interaction is only indirectly involved by means of the quark distributions. This, however, is just an approximation (corresponding to setting $\alpha_s \to 0$) that turns out to be insufficient for more precise experimental investigations. The role of the strong interaction and the impact of QCD on the parton distributions will be the topic of the next section.

5.4.3 *Parton distributions and QCD*

Accurate measurements of the structure functions show small deviations from the scaling behaviour, thus documenting a visible, though weak, dependence on Q^2. This *scaling violation* can be traced back to a Q^2-dependence of the quark distribution functions, $q(x) \to q(x, Q^2)$. The reason is found in the strong interaction, which indeed is suppressed at high Q^2, but nevertheless gives rise to measureable effects that can be described by perturbative QCD.

Starting point is the free-parton scattering process in the $\alpha_s = 0$ approximation, as depicted in Fig. 5.3. Accordingly, the quark with momentum $p = xP$ interacts with the virtual photon with momentum q corresponding to the momentum transfer $Q^2 = -q^2$. At the next order, the lowest order in QCD, the quark–gluon strong interaction induces the emission of a gluon off

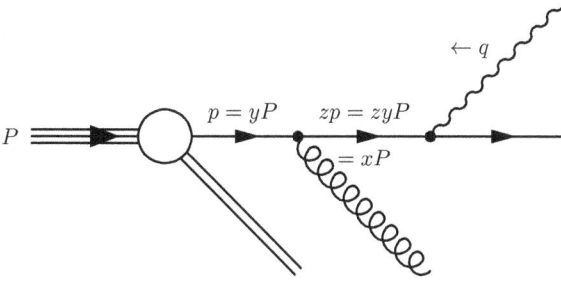

Fig. 5.4 Gluon radiation off quarks in deep-inelastic scattering.

the incoming quark, reducing the quark momentum connected to the QED vertex by the radiated gluon momentum (see Fig. 5.4) whereby the effective quark momentum distribution is different from the original one.

The details of the calculation are lengthy and will be skipped here. As a result, after integrating over the gluon phase space, one obtains a contribution to the eP cross section that can be written as a contribution to the quark distribution function $q(x)$ in Eq. (5.56) as follows,

$$q(x) \to q(x) + \frac{\alpha_s}{2\pi} \log\left(\frac{Q^2}{\mu^2}\right) \int_x^1 \frac{dy}{y}\, q(y) \cdot \underbrace{\frac{4}{3}\frac{1+z^2}{1-z}}_{= P_{qq}(z)}\Bigg|_{z=x/y}.$$

Hence, the original quark distribution becomes Q^2-dependent,

$$q(x) \to q(x) + \frac{\alpha_s}{2\pi} \log\left(\frac{Q^2}{\mu^2}\right) \int_x^1 \frac{dy}{y}\, q(y)\, P_{qq}\left(\frac{x}{y}\right) \equiv q(x, Q^2) \qquad (5.61)$$

and accordingly the structure functions receive a Q^2-dependence as well,

$$F_2(x) \to x \sum_q Q_q^2\, q(x, Q^2) \equiv F_2(x, Q^2). \qquad (5.62)$$

The quantity $P_{qq}(z)$ is the *splitting function*, specifying the probability for a quark with momentum zp to arise from a quark with momentum p by collinear emission of a gluon.

A brief outline of the essential steps leading to Eq. (5.61) shall give some information for a basic understanding how scaling violation comes about.

The matrix element $\mathcal{M}$ for eq scattering with gluon radiation, schematically depicted in Fig. 5.4 together with the notation for the momenta, contains the (massless) quark propagator

$$\frac{1}{(p-k)^2} = -\frac{1}{2kp} = -\frac{1}{2k^0 p^0 \left(1 - \cos\theta^*\right)}$$

with the angle $\theta^* = \angle(\vec{k}, \vec{p})$. For $\theta^* \to 0$, i.e. for a gluon emitted collinear to the quark, a singularity arises in the matrix element and thus in the cross section. This singular collinear part provides the main contribution to the cross section. It can be extracted by isolating the singular term in the squared matrix element and determination of its coefficient in the collinear limit $\theta^* = 0$, after colour and spin summation, according to

$$\overline{|\mathcal{M}|^2} = \frac{1}{k_{\mathrm{T}}^2} \left[\ \cdots \ \right]_{\theta^*=0}$$

where θ^* has been traded for the component $k_{\mathrm{T}} = |\vec{k}| \sin\theta^*$ of the gluon momentum that is transverse to the quark momentum $\vec{p}$. In the collinear limit the momenta can be written as $p - k = zp$ and $k = (1-z)p$ with $0 < z < 1$, and one obtains the singular matrix element squared as follows,

$$\overline{|\mathcal{M}|^2} = \frac{1}{k_{\mathrm{T}}^2} \cdot 2g_s^2 \cdot \frac{4}{3} \left(1 + z^2\right) \cdot \overline{|\mathcal{M}_0|^2} \tag{5.63}$$

where $\mathcal{M}_0$ is the lowest-order matrix element without radiation, given in Eq. (5.52). In the cross section an additional phase space element for the gluon is accrued, which after integration over the azimuth can be written as

$$\frac{1}{(2\pi)^3} \frac{d^3 k}{2k^0} \to \frac{1}{(2\pi)^2} \frac{1}{2k^0} |\vec{k}|^2 \, d|\vec{k}| \, d\cos\theta^* = \frac{1}{16\pi^2} \frac{dk^0}{k^0} dk_{\mathrm{T}}^2 = \frac{1}{16\pi^2} \frac{dz}{1-z} dk_{\mathrm{T}}^2$$

yielding in combination with (5.63)

$$\frac{1}{(2\pi)^3} \frac{d^3 k}{2k^0} \overline{|\mathcal{M}|^2} = \frac{\alpha_s}{2\pi} \frac{dk_{\mathrm{T}}^2}{k_{\mathrm{T}}^2} dz \cdot \frac{4}{3} \frac{1+z^2}{1-z} \cdot \overline{|\mathcal{M}_0|^2}.$$

The integration over the transverse momentum diverges at the lower limit and thus requires a regularization by means of a small mass scale μ as a cut-off parameter; the upper limit is proportional to Q^2 and can be replaced by Q^2,

$$\int_{\mu^2}^{Q^2} \frac{dk_{\mathrm{T}}^2}{k_{\mathrm{T}}^2} = \log \frac{Q^2}{\mu^2}.$$

In this way one obtains the collinear contribution to the differential eq cross section in a factorized form,

$$d\sigma^{eq} = d\sigma_0^{eq} \cdot \frac{\alpha_s}{2\pi} \log\left(\frac{Q^2}{\mu^2}\right) P_{qq}(z) \, dz$$

which contains the lowest-order cross section $d\sigma_0^{eq}$ given in Eq. (5.55). To arrive now at the hadronic eP cross section in terms of the variables x and Q^2 one multiplies $d\sigma^{eq}$ with the parton distribution $q(y)$ of the quark q, refering to the quark momentum $p = yP$, and integrates over all $y > x$ so that after radiation of k the residual momentum $zp = zyP \equiv xP$ is left for the quark. On top, for completing the gluon phase space integration, one has to integrate over z,

$$d\sigma^{eP} = \int_x^1 dy \, q(y) \int_0^1 \delta(yz - x) \, d\sigma^{eq}$$

$$= d\sigma_0^{eq} \int_x^1 \frac{dy}{y} \, q(y) \frac{\alpha_s}{2\pi} \log\left(\frac{Q^2}{\mu^2}\right) P_{qq}\left(\frac{x}{y}\right)$$

$$\equiv d\sigma_0^{eq} \cdot \Delta q(x, Q^2)$$

yielding thus the modification $q(x) \to q(x) + \Delta q(x, Q^2)$ of the quark distribution function as specified in Eq. (5.61).

Note that the original "bare" distributions are not measureable, but only in combination with the QCD correction term (5.61) containing the unphysical cut-off parameter μ. The x-dependence of the quark distributions still has to be determined from measurements of the structure functions. The determination at a given value Q_0^2 provides $q(x, Q_0^2)$ as an experimental input quantity. In the difference

$$q(x, Q^2) - q(x, Q_0^2) = \frac{\alpha_s}{2\pi} \log\left(\frac{Q^2}{Q_0^2}\right) \int_x^1 \frac{dy}{y} \, q(y, Q_0^2) \, P_{qq}\left(\frac{x}{y}\right) \qquad (5.64)$$

the unphysical parameter μ drops out, and in the integrand $q(y)$ has been replaced by $q(y, Q_0^2)$ (the difference is of the order α_s^2). Thus, $q(x, Q^2)$ for general Q^2 follows from $q(x, Q_0^2)$ as a QCD prediction.

Differentiating with respect to Q^2 yields

$$Q^2 \frac{\partial q(x, Q^2)}{\partial Q^2} = \frac{\alpha_s}{2\pi} \int_x^1 \frac{dy}{y} \, q(y, Q_0^2) \, P_{qq}\left(\frac{x}{y}\right) \qquad (5.65)$$

which by the replacement

$$q(y, Q_0^2) \to q(y, Q^2)$$

in the integrand becomes a differential equation for the function q,

$$Q^2 \frac{\partial q(x, Q^2)}{\partial Q^2} = \frac{\alpha_s}{2\pi} \int_x^1 \frac{dy}{y} q(y, Q^2) P_{qq}\left(\frac{x}{y}\right). \qquad (5.66)$$

This equation is of the type of an evolution equation for the Q^2-evolution of the quark distributions, similar to the RGE for the running coupling constant. Also here, a systematic approach exists based on the renormalization group. The splitting function can be determined perturbatively in powers of α_s.

Solving Eq. (5.66) iteratively, starting from the zeroth-order approximation $q^{(0)}(x, Q^2) = q(x, Q_0^2)$, the expression in Eq. (5.64) is obtained as the first approximation, i.e. the approximative solution after the first iteration. Further iterations provide the contributions from multiple radiation in the collinear limit. The complete solution corresponds to the resummation of all successive collinear emissions of gluons.

At order α_s there is another contribution to the quark distribution originating from the gluons in the nucleon: the transition of a gluon with momentum p into a quark–antiquark pair yields a quark (or an antiquark) with momentum zp that interacts with the virtual photon. In the cross section one finds again a singularity $\sim 1/k_T^2$ from collinear quark–antiquark pair production, attributed to a splitting function $P_{qg}(z)$. Performing analogous steps as above for quark–gluon splitting, the following expression is obtained

$$\frac{\alpha_s}{2\pi} \log\left(\frac{Q^2}{\mu^2}\right) \int_x^1 \frac{dy}{y} G(y) P_{qg}\left(\frac{x}{y}\right) \equiv \Delta q(x, Q^2)_{\text{gluon}}$$

as the gluon contribution to the quark distribution function which has to be added in Eq. (5.61). The bare gluon density $G(y)$ is unphysical as well and has to determined in combination with the cutoff μ from measurements at the initial value Q_0^2, yielding $G(y, Q_0^2)$ as an input quantity. On the right side of Eq. (5.64) hence one has to add

$$\frac{\alpha_s}{2\pi} \log\left(\frac{Q^2}{Q_0^2}\right) \int_x^1 \frac{dy}{y} G(y, Q_0^2) P_{qg}\left(\frac{x}{y}\right)$$

and, accordingly, also in the evolution equation (5.66) with $G(y, Q_0^2) \to G(y, Q^2)$.

The gluon distribution itself receives at order α_s contributions from gluon splitting $G \to GG$ with the splitting function P_{gg} and from gluon radiation off quarks as in Fig. 5.4, but now with quark and gluon exchanged, with the splitting function P_{gq}, i.e. the parton momentum zp is now carried by the gluon and the quark with momentum $k = (1-z)p$ is considered a collinearly radiated particle. The gluon can originate from an antiquark as well. The splitting functions are the same for quarks und antiquarks, hence $P_{\bar{q}\bar{q}} = P_{qq}$, $P_{\bar{q}g} = P_{qg}$, $P_{g\bar{q}} = P_{gq}$. The splitting functions are listed at the end of this section.

Because of the various contributions the evolution equations for the quark and gluon distributions become coupled equations (the first one is valid also for $\bar{q}(x, Q^2)$ of antiquarks)

$$Q^2 \frac{\partial q(x, Q^2)}{\partial Q^2} = \frac{\alpha_s}{2\pi} \int_x^1 \frac{dy}{y} \left[P_{qq}\left(\frac{x}{y}\right) q(y, Q^2) + P_{qg}\left(\frac{x}{y}\right) G(y, Q^2) \right],$$

$$Q^2 \frac{\partial G(x, Q^2)}{\partial Q^2} = \frac{\alpha_s}{2\pi} \int_x^1 \frac{dy}{y} \left[P_{gq}\left(\frac{x}{y}\right) \left(q(y, Q^2) + \bar{q}(y, Q^2) \right) \right.$$

$$\left. + P_{gg}\left(\frac{x}{y}\right) G(y, Q^2) \right]. \tag{5.67}$$

They are named DGLAP equations, according to their founders Dokshitzer, Gribov, Lipatov, Altarelli, Parisi [Altarelli (1977); Dokshitzer (1977); Gribov (1972)].

The solutions with initial values at Q_0^2 can be computed either numerically or approximatively by iteration. They can be interpreted as resummation of all successive collinear radiation processes. Examples are graphically displayed in Fig. 5.5.

The x-distributions for quarks, antiquarks, and gluons are shown in Fig. 5.6 for two different values of Q^2 (labeled as μ^2 in the figure). As one can see, at high Q^2, as e.g. at the LHC, gluons are dominant in the small-x region. For $t\bar{t}$ production the relevant x-values are around ~ 0.03, hence gluon fusion is the most important parton subprocess at the LHC. At the Tevatron, instead, with $\sqrt{S} = 2\,\text{TeV}$, values of $x \sim 0.2$ are required,

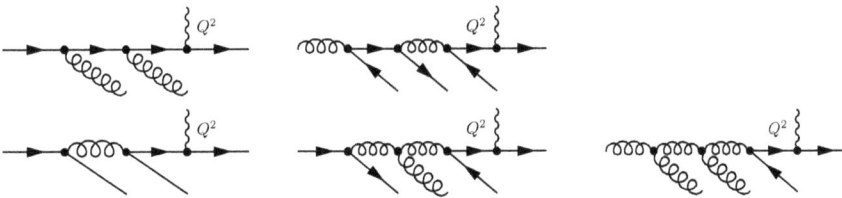

Fig. 5.5 Successive radiation of gluons and quarks/antiquarks.

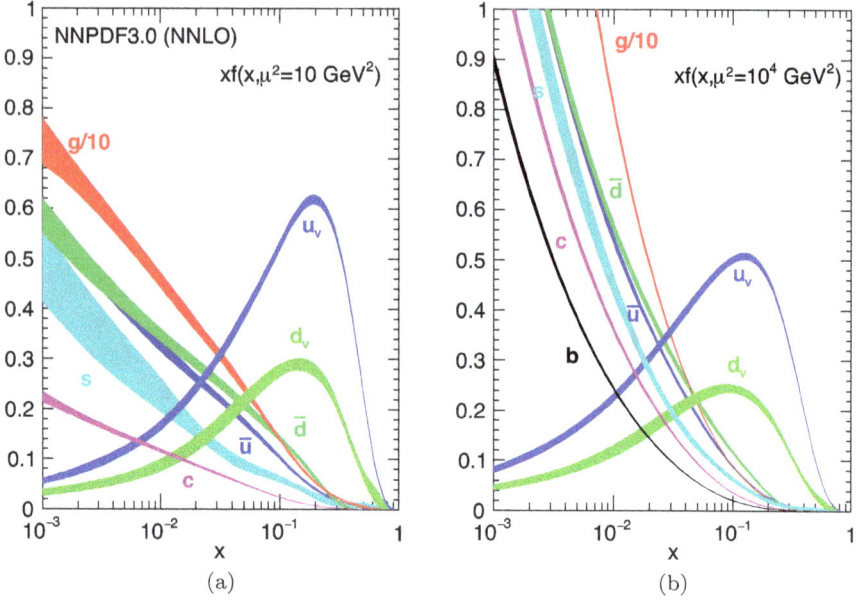

Fig. 5.6 *x*-dependence of the parton distribution functions for valence quarks, sea quarks, and gluons. u_V, d_V denote the valence quarks [Zyla (2020)].

thus in the region where the valence quarks dominate and $q\bar{q} \rightarrow t\bar{t}$ is the most relevant parton channel.

The splitting functions at lowest order $\mathcal{O}(\alpha_s)$ for the various types of radiation are given by the following expressions,

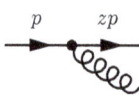
$$P_{qq}(z) = \frac{4}{3} \cdot \frac{1+z^2}{1-z}$$

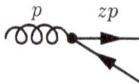
$$P_{qg}(z) = \frac{1}{2} \left[z^2 + (1-z)^2 \right]$$

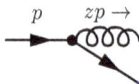
$$P_{gq}(z) = \frac{4}{3} \cdot \frac{1+(1-z)^2}{z}$$

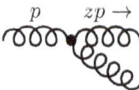
$$P_{gg}(z) = 6 \left[\frac{z}{1-z} + \frac{1-z}{z} + z(1-z) \right]$$

In general, $P_{ab}(z)$ determines the probability for a parton a with momentum zp to originate from a parton b with momentum p.

5.5 Hadronic Bound States

Quarks und antiquarks as constituents of hadronic matter build the mesons as quark–antiquark ($q\bar{q}$) and the baryons as 3-quark (qqq) bound states, with the strong interaction providing the responsible dynamics. QCD is the established fundamental theory of the strong interaction and thus in principle should be able to describe also the binding problem and the hadron spectrum. As already mentioned in Sec. 5.3, specific non-perturbative methods are required in this energy range which cannot be covered here in detail. General statements, on the other hand, like the classification of bound states according to colour and other quantum numbers can be derived directly from the underlying symmetry groups for colour, flavour, CP, as well as for space–time symmetries like rotations and parity.

With respect to colour, hadron states are neutral which means that they appear as singlets among the various possible representations of the group $SU(3)$. The colour content of $q\bar{q}$ and qqq bound states has to be determined from the basic quark and antiquark colour states; the respective composed systems are described mathematically in terms of the product space of the individual one-particle spaces. Therefore, a more detailed discussion is required on the colour states and colour representations of quarks and antiquarks, and of more-particle systems.

5.5.1 *Representations of* $SU(3)$

So far only the fundamental representation for the triplet of quarks was taken into account (and the octet of the adjoint representation). Now also other representations are needed, namely those refering to antiquarks and to composed systems.

- *Quarks in triplet-representation*

  Notation: [3], generators: $T_a = \frac{1}{2}\lambda_a$ $(a = 1, \dots, 8)$

$$\text{special}: \quad T_3 = \frac{1}{2}\begin{pmatrix} 1 & 0 & 0 \\ 0 & -1 & 0 \\ 0 & 0 & 0 \end{pmatrix}, \quad T_8 = \frac{1}{2\sqrt{3}}\begin{pmatrix} 1 & 0 & 0 \\ 0 & 1 & 0 \\ 0 & 0 & -2 \end{pmatrix}.$$

  $[T_3, T_8] = 0 \Rightarrow T_3$ and T_8 have common eigenvectors, which are basis vectors of [3] and thus span the 3-dimensional representation space.

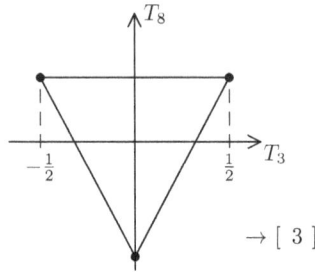

Fig. 5.7 Triplet representation [3] of $SU(3)$.

The basis vectors are determined by the eigenvalues t_3, t_8 of T_3 and T_8. They can be displayed in a T_3–T_8 diagram as the corners of a triangle oriented top-down, as shown in Fig. 5.7.

In the conventional notation of quantum mechanics, eigenvectors are labeled by their simultaneous eigenvalues, hence written in the form $|t_3, t_8\rangle$, or abbreviated as $|$colour$\rangle$, with colour $= R, G, B$ defined as follows,

$$\left| +\frac{1}{2}, \frac{1}{2\sqrt{3}} \right\rangle \equiv |R\rangle$$

$$\left| -\frac{1}{2}, \frac{1}{2\sqrt{3}} \right\rangle \equiv |G\rangle$$

$$\left| 0, -\frac{1}{\sqrt{3}} \right\rangle \equiv |B\rangle$$

- *Antiquarks in anti-triplet representation*

Notation: $[\bar{3}]$,   generators: $\overline{T}_a = -T_a^* = -\frac{1}{2}\lambda_a^*$ $(a = 1, \ldots, 8)$

The generators obey the $SU(3)$ Lie algebra

$$[\overline{T}_a, \overline{T}_b] = i f_{abc} \overline{T}_c$$

and form a representation that is not equivalent to [3]. The generators $\overline{T}_3 = -T_3$ and $\overline{T}_8 = -T_8$ commmute and thus have common eigenvectors refering to the eigenvalues $\bar{t}_3 = -t_3$ and $\bar{t}_8 = -t_8$, which span $[\bar{3}]$ as basis vectors. Short-hand notation: $|$anticolour$\rangle$ with anticolour $= \bar{R}, \bar{G}, \bar{B}$,

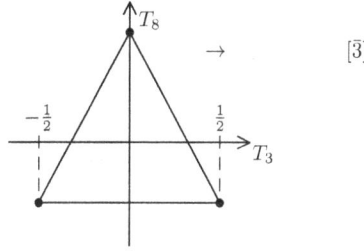

Fig. 5.8 Anti-triplet representation $[\bar{3}]$ of $SU(3)$.

defined as follows,

$$\left| -\frac{1}{2}, -\frac{1}{2\sqrt{3}} \right\rangle \equiv |\bar{R}\rangle$$

$$\left| +\frac{1}{2}, -\frac{1}{2\sqrt{3}} \right\rangle \equiv |\bar{G}\rangle$$

$$\left| 0, +\frac{1}{\sqrt{3}} \right\rangle \equiv |\bar{B}\rangle$$

In a T_3–T_8 diagram they can be displayed as the corners of a triangle oriented bottom-up, as shown in Fig. 5.8.

- *$q\bar{q}$ states in product space of q and $\bar{q}$ spaces*

Product space: $[3] \otimes [\bar{3}]$

Product basis: $|R\rangle|\bar{R}\rangle \equiv |R\bar{R}\rangle, |R\bar{G}\rangle, |R\bar{B}\rangle, |G\bar{R}\rangle, \ldots, |B\bar{B}\rangle$

These 9 basis vectors span the 9-dimensional product space $[3] \otimes [\bar{3}]$. The generators T_a in the product space are represented by 9×9-matrices. The product representation is *reducible*, i.e. by a suitable basis transformation *product basis → new basis* a block-diagonal form is achieved for all $a = 1, \ldots, 8$,

$$T_a = (9 \times 9) \quad \rightarrow \quad \left(\begin{array}{c|c} (8 \times 8) & 0 \\ \hline 0 & (1 \times 1) \end{array} \right) \tag{5.68}$$

The product representation thus can be decomposed into a direct sum of an 8-dimensional representation (octet) and a 1-dimensional representation (singlet),

$$[3] \otimes [\bar{3}] \rightarrow [8] \oplus [1]. \tag{5.69}$$

Octet and singlet are represented by the block matrices in Eq. (5.68). They form the *irreducible representations*, which cannot be further decomposed simultaneously by a basis transfomation. The basis vectors for octet and singlet are linear combinations of the product basis vectors. For the eigenvalues of T_3 and T_8 the following relation holds,

$$t_a^{[3]\otimes[3]} = t_a^{[3]} + t_a^{[3]}, \quad a = 3, 8. \tag{5.70}$$

Analogy:

Coupling of angular momenta, for example

$$\left[\text{spin } \frac{1}{2}\right] \otimes \left[\text{spin } \frac{1}{2}\right] = [\text{spin } 1] \oplus [\text{spin } 0]$$

In general, the transformation from the product basis for spin s_1 and s_2 into the basis of the irreducible representations for total spin S,

$$\underbrace{|s_1 m_1\rangle |s_2 m_2\rangle}_{\text{product basis}} \rightarrow \underbrace{|SM\rangle \quad \text{for } S = |s_1 - s_2|, \ldots, s_1 + s_2}_{\text{new basis}}$$

according to (with $M = m_1 + m_2$)

$$|s_1 m_1\rangle |s_2 m_2\rangle = \sum_S \sum_M C_{m_1 m_2}^{SM} |SM\rangle$$

is done with the Clebsch-Gordan coefficients of the rotation group $SO(3)$, which is algebraically equivalent to $SU(2)$.

For $SU(3)$ analogous Clebsch-Gordan coefficients exist for the transformation from the product basis into the singlet–octet basis. For the colour-singlet one has

$$|\,\text{singlet}\rangle = \frac{1}{\sqrt{3}} \left(|R\bar{R}\rangle + |G\bar{G}\rangle + |B\bar{B}\rangle\right) \tag{5.71}$$

with the eigenvalues $t_3^{[1]} = 0$, $t_8^{[1]} = 0$. The octet basis consists of 8 further linear combinations of the product basis vectors that are orthogonal to (5.71), graphically displayed in Fig. 5.9. They are not colour-neutral.

A graphical method to obtain the states of the product space $[3] \otimes [\bar{3}]$ and the irreducible subspaces contained therein is given by the

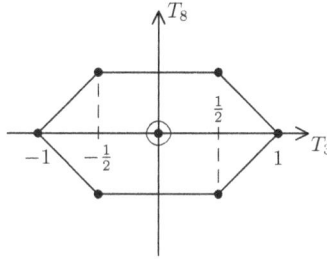

Fig. 5.9 Octet representation [8] of $SU(3)$. The zero is occupied twofold.

following prescription that corresponds to adding the eigenvalues according to Eq. (5.70):

> Put the zero point of the (anti-)triangle of $[\bar{3}]$ on each corner of the triangle of $[3]$. The points resulting from this construction build a hexagon and a triply occupied zero. One of the zero-states corresponds to the singlet, the other two together with the corners of the hexagon provide the states of the octet.

- *qqq states in product space of three q spaces*

Product space: : $[3] \otimes [3] \otimes [3]$

The product representation for qq contains two irreducible representations,

$$[3] \otimes [3] \quad \rightarrow \quad [6] \oplus [\bar{3}]$$

a sextet and an anti-triplet; both are not colour-neutral (and as diquark states they have not been observed so far). In combination with the third quark

$$[3] \otimes [3] \otimes [3] = \left([6] \oplus [\bar{3}]\right) \otimes [3]$$
$$= \underbrace{[6] \otimes [3]}_{} \oplus \underbrace{[\bar{3}] \otimes [3]}_{}$$
$$= [10] \oplus [8] \oplus [8] \oplus [1]$$

one obtains a decuplet, two octets, and a singlet as irreducible representations. Only the singlet is colour-neutral; the corresponding state vector in colour space is given by the totally antisymmetric linear combination of the product basis vectors,

$$|\text{singlet}\rangle = \frac{1}{\sqrt{3!}} \begin{vmatrix} |R\rangle_1 & |G\rangle_1 & |B\rangle_1 \\ |R\rangle_2 & |G\rangle_2 & |B\rangle_2 \\ |R\rangle_3 & |G\rangle_3 & |B\rangle_3 \end{vmatrix}, \qquad (5.72)$$

in analogy to the Slater determinant in quantum mechanics for fermionic more-particle systems.

5.5.2 Hadron states

The total state vector of a meson or a baryon can be composed of colour and other degrees of freedom in the following way,

$$|\text{hadron}\rangle = |\text{colour}\rangle \, |\text{flavour, spin, orbit}\rangle$$

where for the colour content $|\text{colour}\rangle = |\text{singlet}\rangle$ the respective singlet state has to be inserted, either from Eq. (5.71) for mesons or Eq. (5.72) for baryons. The residual parts contain the quark flavours, total spin and orbital angular momentum, as well as the radial part of the wave function.

Accordingly, for the total state vector of a baryonic bound state one has

$$|\text{baryon}\rangle = \underbrace{|\text{singlet}\rangle}_{\text{antisymm.}} |\text{flavour, spin, orbit}\rangle$$

with the antisymmetric singlet (5.72). The non-colour part in total hence has to be symmetric under permutations in order to obey the Pauli principle in case of a baryon consisting only of quarks of the same flavour; the flavour part is automatically symmetric, and thus only symmetric combinations of spin and orbital angular momentum yielding the spin of the baryon are allowed.

Example. Ω^- as a baryon with spin $3/2$ composed of three s-quarks.

$$|\Omega^-\rangle = |\text{colour}\rangle \, |\text{flavour}\rangle \, |\text{spin}\rangle \, |\text{orbit}\rangle$$

$$= |\text{singlet}\rangle \underbrace{|sss\rangle}_{\text{sym}} \underbrace{\left|S = \frac{3}{2}\right\rangle}_{\text{sym}} \underbrace{|L = 0\rangle}_{\text{sym}}$$

The state is symmetric under permutations with respect to the flavour content. The spin part with total spin $S = 3/2$ is symmetric as well, and the total orbital momentum $L = 0$ corresponds to a symmetric wave function either. Only the antisymmetric colour-singlet makes sure that the state vector in total is antisymmetric. Without the colour degree of freedom the Ω^- state would be symmetric and thus in contradiction to the Pauli principle. Historically this was an important indication for the existence of an additional quantum number for the quarks and hence finally for the existence of colour.

Remark. The representations of $SU(3$ are also valid for the flavour-$SU(3)$, which is realized approximatively for the u, d, s quarks and antiquarks and for their bound states. Differently from colour, however, also non-singlet representations of the flavour-$SU(3)$ appear in the hadron spectrum such as the irreducible representations [6], [8], [10], ...

Chapter 6

Electroweak Theory

After the electromagnetic and the strong interaction the third fundamental force, the weak interaction, requires an adequate theoretical description. As it turns out, a consistent fomulation is only possible in combination with QED, which amounts to a unification of the electromagnetic and the weak interaction in terms of a common "electroweak interaction". As in the previous cases, the basic ansatz follows the principles for a description within the framework of a gauge theory, but immediately a central problem occurs: the gauge bosons $W^{\pm}, Z^0$ corresponding to the weak interaction are not massless, as required by gauge symmetry. A new concept is needed in order to reconcile local gauge invariance and massive gauge bosons: the Higgs mechanism. Moreover, also the masses of the charged leptons and quarks are in a conflict with gauge symmetry because of the missing left–right symmetry of the weak interaction. A solution of this conflict is achieved with the help of a Yukawa interaction that has to be introduced in addition. The gauge theory of the electroweak interaction with all these special features is by now commonly accepted and named the *electroweak Standard Model*.

6.1 Historical Overview

Originally the weak interaction was observed by means of particle decays induced by the weak force, such as β-decay of atomic nuclei, μ-decay, π-decays, and many more. Accordingly, the theoretical developments to grip the weak interaction have a long tradition, beginning with Fermi's first ansatz for a Lagrangian to describe β-decay until the completion of the electroweak theory in terms of the electroweak Standard Model.

Table 6.1   Lepton and quark generations.

1st	2nd	3rd	1st	2nd	3rd
ν_e	ν_μ	ν_τ	u	c	t
e	μ	τ	d	s	b

1933   Fermi postulates a pointlike coupling of two currents, with a contact interaction $\mathcal{L}_{\text{int}} \sim J_\rho J^\rho$

1957   Discovery of parity violation, $V - A$ structure of the currents $J_\rho = V_\rho - A_\rho$, vector current V_ρ, axialvector current A_ρ

1968–1973   Formulation of the Standard Model

Experimental verification:

1973   Discovery of neutal currents (Gargamelle bubble chamber)

1983   Discovery of the massive vector bosons $W^\pm, Z^0$ (SPS at CERN)

1989–today   Precision tests of the theory (LEP, SLC, Tevatron, LHC)

1995   Discovery of the top quark (Tevatron)

2012   Discovery of the Higgs boson (LHC)

From a today's point of view, the weak interaction appears as a fundamental interaction of leptons and quarks, which are arranged in generations as shown in Table 6.1, from light to heavy. Each generation consists of a family of leptons and of quarks.

The currents of the weak interaction are formed by the spinor fields of the leptons and of the quarks.

6.2   Chiral Fermions and Chiral Currents

In QED and QCD the interaction terms of fermions contain only vector currents of the type

$$\overline{\psi}\gamma^\mu\psi = V^\mu = (V^0, \vec{V}\,).$$

Under a space inversion **P** a vector current transforms as follows,

$$\mathbf{P}: \quad V^0 \to V^0, \quad \vec{V} \to -\vec{V}.$$

In case of the weak interaction also axialvector currents (or axial currents) occur,

$$\overline{\psi}\gamma^\mu\gamma_5\psi = A^\mu = (A^0, \vec{A}),$$

which get the opposite sign under a space inversion (see Sec. 2.3),

$$\mathbf{P}: \quad A^0 \to -A^0, \quad \vec{A} \to +\vec{A}.$$

6.2.1 *Chirality*

Chirality is defined by the matrix γ_5 as an operator with eigenvalues ± 1, providing an important quantum number of fermions.

- Eigenvalue equation $\quad \gamma_5\psi = \pm\psi$
- Eigenstates

  for $+1$: $\quad \gamma_5\,\psi_R = +\psi_R \quad$ right-chiral or right-handed
  for -1: $\quad \gamma_5\,\psi_L = -\psi_L \quad$ left-chiral or left-handed

- Projectors

$$P_R = \frac{1+\gamma_5}{2}, \quad P_L = \frac{1-\gamma_5}{2} \quad \text{with } P_L^2 = P_L, \quad P_R^2 = P_R, \quad P_R P_L = 0.$$

Each spinor can be decomposed according to $\psi = \psi_L + \psi_R$ into chiral components

$$\psi_L = \frac{1-\gamma_5}{2}\,\psi, \quad \psi_R = \frac{1+\gamma_5}{2}\,\psi. \tag{6.1}$$

Chiral currents are defined as follows,

- left-handed current

$$\overline{\psi}_L\gamma^\mu\psi_L = \overline{\psi}\,\gamma^\mu\frac{1-\gamma_5}{2}\,\psi \equiv J_L^\mu \tag{6.2}$$

- right-handed current

$$\overline{\psi}_R\gamma^\mu\psi_R = \overline{\psi}\,\gamma^\mu\frac{1+\gamma_5}{2}\,\psi \equiv J_R^\mu \tag{6.3}$$

The following relations hold:

$$J_L^\mu + J_R^\mu = V^\mu, \quad -J_L^\mu + J_R^\mu = A^\mu, \quad J_L^\mu = \frac{1}{2}\left(V^\mu - A^\mu\right). \tag{6.4}$$

Chirality is a rather abstract quantity. It becomes more illustrative for relativistic particles with high momentum $|\vec{p}| \gg m$ or for particles with mass $m = 0$. The respective spinor $u(p)$ can be written as follows, with $\vec{n} = \vec{p}/|\vec{p}|$,

$$u(p) = \begin{pmatrix} \varphi \\ (\vec{\sigma}\cdot\vec{n})\varphi \end{pmatrix} \tag{6.5}$$

where φ is a 2-component spinor. Because of $(\vec{\sigma}\cdot\vec{n})^2 = \mathbb{1}$, one obtains

$$\gamma_5\,u(p) = \begin{pmatrix} 0 & \mathbb{1} \\ \mathbb{1} & 0 \end{pmatrix} \begin{pmatrix} \varphi \\ (\vec{\sigma}\cdot\vec{n})\varphi \end{pmatrix} = \begin{pmatrix} \vec{\sigma}\cdot\vec{n} & 0 \\ 0 & \vec{\sigma}\cdot\vec{n} \end{pmatrix} \begin{pmatrix} \varphi \\ (\vec{\sigma}\cdot\vec{n})\varphi \end{pmatrix} \tag{6.6}$$

which implies

$$\gamma_5\,u(p) = \left(\vec{\Sigma}\cdot\vec{n}\right) u(p) = \pm u(p) \tag{6.7}$$

when $u(p)$ is an eigenstate of helicity, hence chirality is (approximatively) equal to helicity. In case of zero mass this equality is exact.

6.3 *V−A* Theory and Massive Vector Bosons

The $V{-}A$ theory, although the name is suggestive, is actually not a genuine theory, but instead is an effective description of processes directed by the weak interaction at low energies. The original ansatz dates back to Fermi; it is based on the pointlike interaction of two currents modified later to $V{-}A$ currents after the discovery of parity violation in weak decay processes,

$$\mathcal{L}_{\text{int}} = \frac{G_F}{\sqrt{2}}\, J_\rho J^\rho \quad \text{with } J_\rho = V_\rho - A_\rho \tag{6.8}$$

involving a universal coupling constant, the *Fermi constant* [Zyla (2020)]

$$G_F = 1.1663787(6) \cdot 10^{-5}\,\text{GeV}^{-2} \tag{6.9}$$

which is not dimensionless. The currents are formed by the spinor fields of the leptons and quarks (for each family seperately).

6.3.1 *Formulation for leptons*

From now on the notation for the Dirac fields of the leptons $e, \mu, \tau, \nu_e, \nu_\mu, \nu_\tau$ will be kept under the following terminology,

$$e(x) = \psi_e(x), \quad \nu_e(x) = \psi_{\nu_e}(x), \quad \mu(x) = \psi_\mu(x), \quad \dots$$

Hence, the leptonic current of the type $V{-}A$ for one family ($\ell = e, \mu, \tau$) is written as follows,

$$J_\rho^{(\ell)} = \bar{\nu}_\ell\, \gamma_\rho(1 - \gamma_5)\, \ell + \bar{\ell}\, \gamma_\rho(1 - \gamma_5)\, \nu_\ell. \tag{6.10}$$

Matrix elements of this current are different from zero only between particle states with different electromagnetic charges, as for example

$$\langle \nu_\ell \,|\, J_\rho^{(\ell)} \,|\, \ell^- \rangle \neq 0, \quad \langle \ell^- \,|\, J_\rho^{(\ell)} \,|\, \nu_\ell \rangle \neq 0.$$

For this reason the current is called a *charged current*.

For leptonic processes due to the weak interaction the S-matrix element (lowest order), by means of $\mathcal{H}_{\text{int}} = -\mathcal{L}_{\text{int}}$, is given by

$$S_{fi} = i \int d^4x \, \langle f \,|\, \mathcal{L}_{\text{int}} \,|\, i \,\rangle = i\, \frac{G_F}{\sqrt{2}} \int d^4x \, \langle f \,|\, J_\rho^{(\ell)}(x) J^{(\ell')\rho}(x) \,|\, i \,\rangle$$

$$= i\,(2\pi)^4\, \delta^4(P_f - P_i)\, \mathcal{T}_{fi} \tag{6.11}$$

where $\mathcal{T}_{fi} = (2\pi)^{-6}\, \mathcal{M}_{fi}$ for processes involving four particles, either decay processes of the type $1 \to 3$ or scattering processes of the type $2 \to 2$.

6.3.2 *Muon decay in the Fermi model*

The muon was discovered already in 1937 and is thus one of the first and best known elementary particles, with a mass of $m_\mu = 105.6583745(24)$ MeV [Zyla (2020)]. Its lifetime is determined by the decay

$$\mu^- \to \nu_\mu + e^- + \bar{\nu}_e \quad \text{or} \quad \mu^+ \to \bar{\nu}_\mu + e^+ + \nu_e$$

mediated by the weak interaction. Experimentally the muon lifetime is known very precisely [Zyla (2020)],

$$\tau_\mu^{\text{exp}} = 2.196\,981\,1\,(22)\,\mu\text{s}.$$

This allows an accurate determination of the Fermi constant by comparison with the theory prediction for the lifetime (natural units)

$$\tau_\mu = \frac{1}{\Gamma_\mu} \tag{6.12}$$

calculated from the decay width Γ_μ. Because of the basic importance for the weak interaction the steps of the calculation will be elucidated in some detail.

For the computation of the width the matrix element for the decay is required. It is obtained from Eq. (6.11) by specifying $\ell = \mu$ and $\ell' = e$ and inserting the states $|i\rangle = |\mu^-, p\rangle$ and $|f\rangle = |\nu_\mu, k\rangle |e^-, p'\rangle |\bar{\nu}_e, k'\rangle$. The result for $\mathcal{M}$ $(=\mathcal{M}_{fi})$ corresponds to the calculation from the Feynman diagram

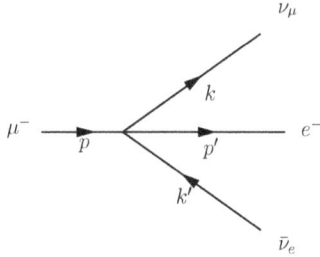

with the usual Feyman rules for the handling of spinors when the 4-point vertex is defined in terms of the Fermi constant. One obtains the expression

$$\mathcal{M} = i\frac{G_F}{\sqrt{2}} \left[\bar{u}(k)\gamma^\rho(1-\gamma_5)u(p)\right] \left[\bar{u}(p')\gamma_\rho(1-\gamma_5)v(k')\right]. \tag{6.13}$$

From the general formula (3.82) the differential decay width is obtained,

$$d\Gamma_\mu = \frac{(2\pi)^{-5}}{2m_\mu} \overline{|\mathcal{M}|^2}\, d\Phi \tag{6.14}$$

involving the phase space element

$$d\Phi = \frac{d^3k}{2k^0}\frac{d^3k'}{2k'^0}\frac{d^3p'}{2p'^0}\, \delta^4(p'+k'+k-p). \tag{6.15}$$

The further calculation is performed in the rest frame of the muon, neglecting the electron mass. With this simplification the spin-averaged matrix element squared reads as follows,

$$\overline{|\mathcal{M}|^2} = 32\, G_F^2\, m_\mu^2\, k'^0\big(m_\mu - 2k'^0\big) \tag{6.16}$$

being independent of the angles. Thus, the phase space integration over the angles can be done seperately in (6.15), yielding

$$d\Phi = \pi^2\, dk'^0 dp'^0 \tag{6.17}$$

and one finds

$$d\Gamma_\mu = \frac{G_F^2\, m_\mu^2}{2\pi^3}\, k'^0\big(m_\mu - 2k'^0\big)\, dk'^0 dp'^0. \tag{6.18}$$

By integration over k'^0 between the kinematical limits $\frac{m_\mu}{2} - p'^0$ and $\frac{m_\mu}{2}$ one obtains the electron spectrum

$$d\Gamma_\mu = \frac{G_F^2\, m_\mu}{12\pi^3}\, (p'^0)^2\, (3m_\mu - 4p'^0)\, dp'^0 \tag{6.19}$$

and the last integration over p'^0 from 0 up to the maximum energy $\frac{m_\mu}{2}$ finally yields the decay width of the muon

$$\boxed{\Gamma_\mu = \frac{G_F^2\, m_\mu^5}{192\pi^3}} \tag{6.20}$$

and accordingly by means of Eq. (6.12) the lifetime τ_μ.

For those interested in the derivation some more details are given for the various steps of the computation.

• Calculation of $\overline{|\mathcal{M}|^2}$

With the help of the trace techniques of Sec. 3.7 one obtains from Eq. (6.13)

$$\overline{|\mathcal{M}|^2} = G_F^2\, \text{Tr}\big[(1-\gamma_5)\, \slashed{k}\,\gamma^\rho\slashed{p}\,\gamma^\sigma\big]\, \text{Tr}\big[(1-\gamma_5)\, \slashed{p}\,'\gamma_\rho\, \slashed{k}\,'\gamma_\sigma\big].$$

A new feature is the trace with an additional γ_5. Supplementary to the list in Sec. 3.7 a further rule on traces over Dirac matrices is required, reading

$$\text{Tr}\big[\gamma_5\gamma^\nu\gamma^\rho\gamma^\lambda\gamma^\sigma\big] = -4i\,\epsilon^{\nu\rho\lambda\sigma} \tag{6.21}$$

with the 4-dimensional ϵ-tensor, obeying the usual convention $\epsilon^{0123} = +1$. Furthermore, one needs the contraction

$$\epsilon^{\rho\sigma\nu\lambda}\epsilon_{\rho\sigma\nu'\lambda'} = -2\left(g^\nu_{\nu'}\, g^\lambda_{\lambda'} - g^\nu_{\lambda'}\, g^\lambda_{\nu'}\right). \tag{6.22}$$

Altogether, one obtains

$$\overline{|\mathcal{M}|^2} = G_F^2 \cdot 64\, (kp')\, (pk').$$

In the muon rest frame with $p = (m_\mu, \vec{0}\,)$ one finds

$$pk' = m_\mu k'^0, \quad 2kp' = (k+p')^2 = (p-k')^2 = m_\mu^2 - 2m_\mu k'^0$$

and thus the expression (6.16) is recovered.

• Calculation of the phase space integral

The phase space element $d\Phi$ in Eq. (6.15) can be written as follows,

$$d\Phi = \delta^4(p' + k' + k - p)\, d^4k\, \delta(k^2)\, \frac{d^3k'}{2k'^0}\, \frac{d^3p'}{2p'^0}.$$

Integrating over k yields

$$d\Phi = \frac{1}{4}\, \delta\big[(k'+p'-p)^2\big]\, d\Omega_{k'}\, d\Omega_{p'}\, k'^0 dk'^0\, p'^0 dp'^0.$$

Choosing the $\vec{p}\,'$ direction as the polar axis for $d\Omega_{k'}$ and integrating over $\varphi_{k'}$ and $\cos\theta_{k'}$, one obtains

$$d\Phi = \frac{1}{4}\, k'^0\, p'^0\, \delta\left[2k'^0 p'^0\,(\cos\theta_{k'} - 1) + 2m_\mu(k'^0 + p'^0) - m_\mu^2\right]$$

$$\times\, d\cos\theta_{k'}\, d\varphi_{k'}\, d\Omega_{p'}$$

$$= \frac{\pi}{2}\, \frac{1}{2k'^0 p'^0}\, d\Omega_{p'}\, k'^0 dk'^0\, p'^0 dp'^0 = \frac{\pi}{4}\, d\Omega_{p'}\, dk'^0\, dp'^0.$$

The final integration over $d\Omega_{p'}$ yields another factor 4π and one ends up with

$$d\Phi = \pi^2\, dk'^0 dp'^0$$

which is the result listed in Eq. (6.17).

A more accurate calculation of the decay width without neglecting the electron mass modifies the expression (6.20) by a correction factor (shown in brackets)

$$\Gamma_\mu = \frac{G_F^2\, m_\mu^5}{192\pi^3}\left(1 - 8x + 8x^3 - x^4 - 12x^2\ln x\right), \qquad x = \frac{m_e^2}{m_\mu^2}, \qquad (6.23)$$

which is very close to unity (0.99981295), but necessary for the determination of the Fermi constant from the muon lifetime in view of the high experimental precision.

6.3.3 *Intermediate vector bosons*

The pointlike current–current coupling finds its explanation by the model of intermediate heavy vector bosons: there is a pair of charged spin-1 particles $W^\pm$ with electromagnetic charge ± 1 (in units of e) and mass $M_W \gg m_\mu$, which couple to the charged current. Their exchange between two currents mediates the interaction between the currents. The $W^\pm$ are described by vector fields $W_\rho^\pm(x)$, as quantum fields with $W_\rho^- = (W_\rho^+)^\dagger$ represented by

$$W_\rho^+(x) = \frac{1}{(2\pi)^{3/2}}\int\frac{d^3k}{2k^0}\sum_{\lambda=\pm,0}\left[a_\lambda(k)\,\epsilon_\rho^\lambda\, e^{-ikx} + b_\lambda^\dagger(k)\,\epsilon_\rho^{\lambda*}\, e^{ikx}\right] \qquad (6.24)$$

with the creation/annihilation operators $a_\lambda^\dagger, a_\lambda$ of W^+ and $b_\lambda^\dagger, b_\lambda$ of W^-, and with polarization vectors ϵ_ρ^λ. Differently from the photon, due to the finite mass $M_W \neq 0$, not only the transverse polarizations refering to helicities $\lambda = \pm 1$ exist but also a longitudinal polarization refering to helicity

$\lambda = 0$. The W fields couple to the left-handed charged current in the following way valid for each family (CC stands for *charged current*),

$$\mathcal{L}_{\text{int}}^{(CC)} = \frac{g}{\sqrt{2}} \left[\bar{\nu}_\ell \gamma^\rho \frac{1-\gamma_5}{2} \ell \, W_\rho^+ + \bar{\ell} \gamma^\rho \frac{1-\gamma_5}{2} \nu_\ell \, W_\rho^- \right] \tag{6.25}$$

with a dimensionless universal coupling constant g. The interaction terms are displayed graphically by vertices supplementing the list of Feynman rules specified so far for QED and QCD:

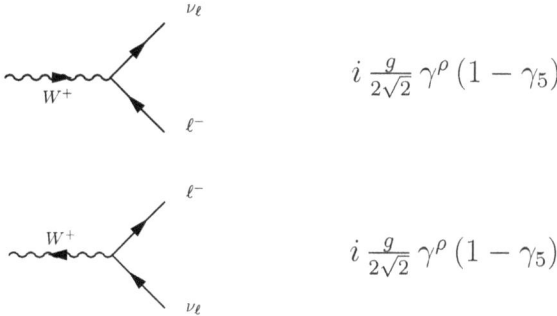

$$i \frac{g}{2\sqrt{2}} \gamma^\rho (1 - \gamma_5)$$

$$i \frac{g}{2\sqrt{2}} \gamma^\rho (1 - \gamma_5)$$

The arrow at the wavy line shows the direction of the flow of the indicated charge. Equivalently to W^+, also W^- with the opposite arrow can be chosen.

In order to describe virtual W bosons as exchange particles the propagator of massive spin-1 fields is required. As usual the propagator is obtained as the causal Green function of the respective field equation. For massive vector fields the field equation is the Proca equation (4.13); the Green function of the W field is thus given by the causal solution of

$$\left[(\Box + M_W^2) g^{\nu\rho} - \partial^\nu \partial^\rho \right]_{(x)} D_{\rho\sigma}(x - y) = g^\nu_\sigma \, \delta^4(x - y).$$

In momentum space, this turns into the algebraic equation

$$\left[(-q^2 + M_W^2) g^{\nu\rho} + q^\nu q^\rho \right] D_{\rho\sigma}(q) = g^\nu_\sigma$$

with the solution (including the $i\varepsilon$ term for the causal boundary condition)

$$\boxed{D_{\rho\sigma}(q) = \frac{1}{q^2 - M_W^2 + i\varepsilon} \left(-g_{\rho\sigma} + \frac{q_\rho q_\sigma}{M_W^2} \right).} \tag{6.26}$$

The graphical symbol for the W propagator (6.26) in the Feynman rules

$$iD_{\rho\sigma}(q)$$

contains an arrow indicating the direction of the flow of the charge, either of W^+ or of W^-. The mathematical expression, however, is independent of the sign of the charge flow (accordingly, the arrow is often omitted in Feynman graphs).

Muon decay. In the model of intermediate vector bosons the matrix element for the decay of the μ^- is represented by a Feynman diagram with a W propagator between two charged-current vertices,

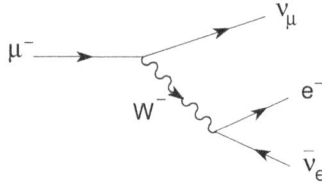

reading, with the new Feynman rules and with the momenta as in Eq. (6.13),

$$\mathcal{M} = i\,\frac{g^2}{8}\,\left[\bar{u}(k)\gamma^\rho(1-\gamma_5)u(p)\right]D_{\rho\sigma}(Q)\left[\bar{u}(p')\gamma^\sigma(1-\gamma_5)v(k')\right]$$

with $Q = p - k$. The Q-dependent term in the numerator of the propagator yields a contribution proportional to $m_\mu m_e / M_W^2$, which due to $M_W \gg m_\mu$ is strongly suppressed and thus negligible. In the denominator, Q^2 is of the order m_μ^2 and hence negligible versus M_W^2 as well. Therefore, the propagator becomes effectively a constant and the matrix element corresponds to that of a pointlike interaction,

$$\mathcal{M} = -i\,\frac{g^2}{8M_W^2}\,\left[\bar{u}(k)\gamma^\rho(1-\gamma_5)u(p)\right]\left[\bar{u}(p')\gamma_\rho(1-\gamma_5)v(k')\right].$$

Comparison with Eq. (6.13) provides an identification of the effective couplings,

$$\boxed{\frac{g^2}{8M_W^2} = \frac{G_F}{\sqrt{2}}}\tag{6.27}$$

from which the coupling constant g can be determined. With $M_W \approx 80\,\mathrm{GeV}$ one obtains

$$\frac{g^2}{4\pi} \simeq \frac{1}{29.8},$$

which is of the same order of magnitude as the fine structure constant α of the electromagnetic interaction.

Neutrino-electron scattering. Another application is found in the elastic scattering of muon-neutrinos by electrons, described by the following Feynman graph

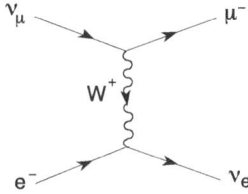

corresponding to the matrix element $\mathcal{M}$, where the approximation in the W-propagator for the low-energy range is taken once again. In performing the spin summation in $\overline{|\mathcal{M}|^2}$ one has to take into account that the neutrino occurs exclusively with helicity $-1/2$ (according to the left-handed chirality) and thus the weight factor $1/2$ has to be dropped. By use of Eq. (6.27) one obtains

$$\overline{|\mathcal{M}|^2} = 16\, G_F^2\, s^2$$

and accordingly the differential cross section in the CMS, with $s = E_{\text{CMS}}^2$,

$$\frac{d\sigma}{d\Omega} = \frac{1}{64\pi^2 s}\,\overline{|\mathcal{M}|^2} = \frac{G_F^2}{4\pi^2}s \tag{6.28}$$

as well as the integrated cross section,

$$\sigma(\nu_\mu e^-) = \frac{G_F^2}{\pi}s. \tag{6.29}$$

The cross section is very small. Evaluating s in the laboratory frame, $s = 2m_e E_\nu$, one finds

$$\sigma(\nu_\mu e^-) = 17{\cdot}10^{-42}\,\text{cm}^2\cdot\frac{E_\nu}{\text{GeV}} \tag{6.30}$$

[making use of the conversion rule $1\,\text{GeV}^{-2} \to 3.92{\cdot}10^{-28}\,\text{cm}^2$ from natural units to SI units]. Neutrino energies E_ν are availabe up to several 100 GeV at high-energy accelerators, by production of π^+ mesons and their subsequent dominant decay $\pi^+ \to \mu^+\,\nu_\mu$.

6.3.4 *Extension to quarks*

The $V{-}A$ interaction of the leptonic charged current with the vector bosons can be transfered to the doublets of quarks retaining the same structure, for the quarks of the first generation given by

$$\mathcal{L}_{\text{int}}^{(CC)} = \frac{g}{\sqrt{2}} \left[\bar{u}\,\gamma^\rho \frac{1-\gamma_5}{2}\, d\; W_\rho^+ + \bar{d}\,\gamma^\rho \frac{1-\gamma_5}{2}\, u\; W_\rho^- \right] \tag{6.31}$$

and with the substitution $u \to c, t$ and $d \to s, b$ for the subsequent generations as well, always with the same universal coupling constant. The vertices listed above for leptons are valid for quarks as well. There is, however, one specific peculiarity for quarks, namely family mixing which allows transitions between quarks also belonging to different generations. This mixing is neglected in Eq. (6.31); it will be treated later in the context of the Standard Model.

Applications concern, for example, weak decays of mesons and baryons where always one of the constituent quarks is converted into its partner quark. In particular one has to mention the β-decay of the neutron $N \to P\,e^-\bar{\nu}_e$ and decays of pions such as $\pi^- \to e^-\bar{\nu}_e$, $\pi^- \to \mu^-\bar{\nu}_\mu$, $\pi^- \to \pi^0\,e^-\bar{\nu}_e$, that have played an important role in the history of the weak interaction.

Neutrino–nucleon scattering. A special role is also played by *deep-inelastic neutrino scattering*, which besides deep-inelastic electron scattering provides essential contributions for the experimental determination of the quark distribution functions. The elementary parton processes consist in scattering of the neutrinos by the quarks of the nucleon, $\nu_\ell\, q \to \ell^- q'$, in analogy to $\nu_\mu\, e^-$ scattering, with the replacement $e^- \to d\,(\bar{u})$ und $\nu_e \to u\,(\bar{d})$. Since there is no difference between ν_e and ν_μ in the scattering process the simpler notation ν is kept for both species.

Due to charge conservation, only the processes

$$\nu\, q \to \ell^-\, q', \quad \nu\,\overline{q'} \to \ell^-\,\bar{q} \quad \text{with } q = d, s, \; q' = u, c$$

are possible. For $\nu\, q \to \ell^-\, q'$ the matrix element is equal to that of $\nu_\mu\, e^- \to e^-\nu_e$ and thus the differential cross section is the same as in Eq. (6.28). It is convenient to rewrite the cross section in terms of the kinematical variable t,

$$\frac{d\sigma}{dt}(\nu q) = \frac{1}{16\pi s^2}\,\overline{|\mathcal{M}|^2} = \frac{G_F^2}{\pi}.$$

For scattering by antiquarks, $\nu\,\overline{q'} \to \ell^-\,\bar{q}$, the result is different,

$$\frac{d\sigma}{dt}(\nu\overline{q'}) = \frac{G_F^2}{\pi}\left(1 + \frac{t}{s}\right)^2$$

with the momentum transfer $Q^2 = -t$. Moreover, it is convenient to use the variable $y = -t/s = Q^2/s$. The differential cross sections for neutrino scattering by proton and neutron hence can be written in the following way, with the quark distributions according to Eqs. (5.49) and (5.50),

$$\frac{d^2\sigma}{dx\,dQ^2}(\nu P) = \frac{G_F^2}{\pi}\left[d(x) + s(x) + (1-y)^2\left[\bar{u}(x) + \bar{c}(x)\right]\right]$$

$$\frac{d^2\sigma}{dx\,dQ^2}(\nu N) = \frac{G_F^2}{\pi}\left[u(x) + s(x) + (1-y)^2\left[\bar{d}(x) + \bar{c}(x)\right]\right].$$

(6.32)

For completeness the cross sections for antineutrino scattering are given as well,

$$\frac{d^2\sigma}{dx\,dQ^2}(\bar{\nu}P) = \frac{G_F^2}{\pi}\left[[u(x) + c(x)](1-y)^2 + \bar{d}(x) + \bar{s}(x)\right]$$

$$\frac{d^2\sigma}{dx\,dQ^2}(\bar{\nu}N) = \frac{G_F^2}{\pi}\left[[d(x) + c(x)](1-y)^2 + \bar{u}(x) + \bar{s}(x)\right].$$

(6.33)

Measurements provide combinations of quark distribution functions that are different from those extracted from electron–nucleon scattering, thus yielding complementary information to distentangle the various quark flavours.

6.3.5 Deficiencies of the V–A theory

In spite of the successful description of weak interaction processes in the low-energy regime, the V–A theory with charged vector bosons proves insufficient in several respects. The problems are based essentially on the existence of a longitudinal polarization state of the $W^\pm$ bosons corresponding to helicity $\lambda = 0$, owing to their finite mass. The polarization vectors ϵ_ρ^λ in Eq. (6.24) refering to helicities $\lambda=0, \pm 1$ are orthogonal to the 4-momentum k and yield the polarization sum

$$\sum_\lambda \epsilon_\rho^\lambda \epsilon_\sigma^{\lambda*} = -g_{\rho\sigma} + \frac{k_\rho k_\sigma}{M_W^2},$$

(6.34)

which is the projector on the subspace orthogonal to k. The transverse polarization vectors corresponding to $\lambda = \pm 1$ are the same as in the case of massless photons. New is the longitudinal polarization vector $\epsilon_\rho^0 \equiv \epsilon_{L\rho}$ related to $\lambda = 0$. It is determined by the properties

$$\epsilon_L \cdot k = 0, \quad \epsilon_L \cdot \epsilon^\pm = 0, \quad \epsilon_L \cdot \epsilon_L = -1,$$

yielding the representation

$$\left(\epsilon_L^\rho\right) = \frac{1}{M_W}\left(|\vec{k}|, k^0\,\vec{n}\right), \quad \vec{n} = \frac{\vec{k}}{|\vec{k}|}.$$

(6.35)

The most remarkable feature of ϵ_L is the high-energy limit $|\vec{k}| \gg M_W$ where owing to $|\vec{k}| \simeq k^0$ the following relation holds,

$$\left(\epsilon_L^\rho\right) \approx \frac{1}{M_W}\left(k^0, \vec{k}\right) = \frac{1}{M_W}\left(k^\rho\right). \tag{6.36}$$

This implies an increase with the momentum, which has a significant impact on the high-energy behaviour of scattering amplitudes and cross sections for processes with external W bosons.

W-pair production in electron–positron annihilation. A particularly instructive example is e^+e^- annihilation into W^+W^- pairs

$$e^+(q) + e^-(p) \rightarrow W^+(k_+) + W^-(k_-)$$

at high energies, $p^0, q^0 \gg M_W$. The following Feynman diagram with neutrino exchange

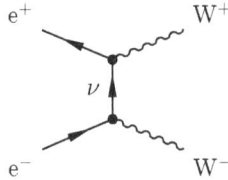

describes the matrix element $\mathcal{M}$. At high energies the longitudinal W-polarizations with $\epsilon_L(k_\pm) \simeq k_\pm/M_W$ dominate and one obtains

$$\mathcal{M} = -i\,\frac{g^2}{8}\left[\bar{v}(q)\gamma_\rho(1-\gamma_5)\,\frac{\not{p}-\not{k}_-}{(p-k_-)^2}\,\gamma_\sigma(1-\gamma_5)u(p)\right]\epsilon_L^\rho(k_+)\,\epsilon_L^\sigma(k_-)$$

$$= i\,\frac{g^2}{4M_W^2}\left[\bar{v}(q)\not{k}_+\left(\not{p}-\not{k}_-\right)\not{k}_-(1-\gamma_5)\,u(p)\right]\frac{1}{(2pk_-)}$$

$$= i\,\frac{g^2}{4M_W^2}\,\bar{v}(q)\not{k}_+(1-\gamma_5)\,u(p) + \cdots$$

where only the leading terms for high energies are kept. Continuing in the same approximation, the standard steps of the calculation yield

$$\overline{|\mathcal{M}|^2} = \left(\frac{g^2}{4M_W^2}\right)^2\cdot\frac{1}{2}\,\mathrm{Tr}\big\lfloor\not{q}\,\not{k}_+\,\not{p}\,\not{k}_+(1-\gamma_5)\big]=\left(\frac{g^2}{4M_W^2}\right)^2\cdot 4\,(pk_+)(qk_+)$$

$$= 2\,G_F^2\cdot\frac{s^2}{4}\,(1-\cos^2\theta)$$

expressed in the CMS with the scattering angle θ, $s = (p+q)^2 = 4E^2$, $E = p^0 = q^0$, and using the relation (6.27). The differential cross section in the CMS thus reads as follows,

$$\frac{d\sigma}{d\Omega} = \frac{1}{64\pi^2 s}\overline{|\mathcal{M}|^2} = \frac{G_F^2}{128\pi^2} s\,(1 - \cos^2\theta)$$

and the integrated cross section becomes

$$\sigma = \frac{G_F^2}{128\pi^2} s \int d\Omega\,(1 - \cos^2\theta) = \frac{G_F^2}{48\pi} s \tag{6.37}$$

pointing out an unlimited increase with $s=4E^2$. Such a behaviour is unacceptable for physics reasons since it means that the fundamental unitarity of the S-matrix, $S^\dagger S = \mathbb{1}$, is violated. The unitarity of the S-matrix restricts the probability for each individual reaction channel, making sure that the sum over the probabilities for all possible final states is unity.

The obvious incompleteness of the $V-A$ theory becomes even more pronounced when the electromagnetic properties of the W bosons are taken into account. Being charged particles, $W^\pm$ have besides the weak interaction with fermions also an electromagnetic interaction with the photon. This has the consequence that also photon exchange in $e^+e^- \to W^+W^-$,

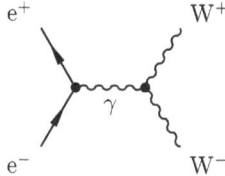

should provide a contribution to the matrix element, proportional to e^2. An ansatz can be made for the γWW vertex that allows all possible Lorentz covariants. None of the various possibilities, however, can compensate the increase of the cross section with energy. Hence, there is clear hint on the close connection between weak and electromagnetic interactions, but also on a decisive missing link.

Algebra of weak currents. An important formal indication towards a potential amendment to the description of the weak interaction via the charged current of leptons and quarks arises from an algebraic point of view.

The CC interaction for leptons and quarks can be written in the following way,

$$\mathcal{L}_{\text{int}}^{(CC)} = \frac{g}{\sqrt{2}} \left(\bar{\nu}_\ell, \bar{\ell}\right) \gamma^\rho \frac{1 - \gamma_5}{2} \left[I_+ W_\rho^+ + I_- W_\rho^-\right] \begin{pmatrix} \nu_\ell \\ \ell \end{pmatrix}$$
$$+ \frac{g}{\sqrt{2}} \left(\bar{u}, \bar{d}\right) \gamma^\rho \frac{1 - \gamma_5}{2} \left[I_+ W_\rho^+ + I_- W_\rho^-\right] \begin{pmatrix} u \\ d \end{pmatrix} \tag{6.38}$$

with the help of the matrices

$$I_+ = \begin{pmatrix} 0 & 1 \\ 0 & 0 \end{pmatrix}, \quad I_- = \begin{pmatrix} 0 & 0 \\ 1 & 0 \end{pmatrix}$$

which are known as the shift operators from the algebra of angular momentum for spin $1/2$,

$$I_\pm = I_1 \pm i I_2, \quad I_a = \frac{1}{2}\sigma_a \quad (a = 1, 2, 3). \tag{6.39}$$

The I_a obey the Lie algebra (4.89) of $SU(2)$; equivalent are the commutator relations

$$[I_+, I_-] = 2I_0, \quad [I_0, I_\pm] = \pm I_\pm \tag{6.40}$$

with $I_0 \equiv I_3$. For a complete realization of this algebra in $\mathcal{L}_{\text{int}}$ an additional current related to I_0 and an associated neutral vector boson W^0 is needed.

6.4 Electroweak Standard Model

The formulation of the Standard Model [Glashow (1961); Weinberg (1967); Salam (1968); Glashow (1970); Kobayashi (1973)] is done in the framework of a non-Abelian gauge theory performing the steps that have been presented in Sec. 4.4: global gauge invariance of a suitable $\mathcal{L}_0$ for the free fermions, local gauge invariance by means of a covariant derivative with gauge fields, and a kinetic term for the intrinsic dynamics of the gauge fields. The problem that the gauge bosons are not massless will be discussed and solved afterwards.

6.4.1 *Symmetry group and global gauge invariance*

The occurrence of leptons and quarks as doublets with respect to the weak interaction leads to the group $SU(2)$, with the fermions classified in the fundamental representation. The generators are given by $T_a \equiv I_a = \frac{1}{2}\sigma_a$, labeled as *isospin* due to the formal analogy to spin. The electric charge

Q is different within a doublet, hence a doublet cannot be a representation of the electromagnetic gauge group. There is, however, a further quantum number valid for an entire doublet: the hypercharge Y, defined by the *Gell-Mann–Nishijima relation*

$$Q = I_3 + \frac{Y}{2}.$$ (6.41)

Each doublet is an eigenstate of Y,

$$Y \begin{pmatrix} \nu \\ e \end{pmatrix} = (-1) \begin{pmatrix} \nu \\ e \end{pmatrix}, \quad Y \begin{pmatrix} u \\ d \end{pmatrix} = \frac{1}{3} \begin{pmatrix} u \\ d \end{pmatrix}.$$ (6.42)

Y is the generator of a group $U(1)$ that has to be distinguished from the electromagnetic $U(1)_{\text{em}}$. The four generators I_a, Y form the Lie algebra

$$\left[I_a, I_b \right] = i\epsilon_{abc} I_c, \quad \left[I_a, Y \right] = 0.$$ (6.43)

This is the Lie algebra of a direct product of two groups, namely

$$\boxed{SU(2) \times U(1)}$$ (6.44)

forming the gauge group of the electroweak interaction.

A special feature of the representations of leptons and quarks is their chirality. The doublet representation of $SU(2)$ contains only left-handed fields, whereas right-handed fields are singlets under $SU(2)$, without right-handed neutrinos. Arranged in families one has the following classification,

$$\begin{pmatrix} \nu_e \\ e \end{pmatrix}_L, \quad \begin{pmatrix} \nu_\mu \\ \mu \end{pmatrix}_L, \quad \begin{pmatrix} \nu_\tau \\ \tau \end{pmatrix}_L, \quad e_R, \quad \mu_R, \quad \tau_R$$

$$\begin{pmatrix} u \\ d \end{pmatrix}_L, \quad \begin{pmatrix} c \\ s \end{pmatrix}_L, \quad \begin{pmatrix} t \\ b \end{pmatrix}_L, \quad u_R, d_R, \quad c_R, s_R, \quad t_R, b_R$$ (6.45)

The respective quantum numbers I_3, Y and charge Q are put together in Table 6.2 for the first generation.

Table 6.2 Quantum numbers for the left- and right-handed leptons and quarks (next generations are replications).

	ν_L	e_L	e_R	u_L	d_L	u_R	d_R
I_3	+1/2	−1/2	0	+1/2	−1/2	0	0
Y	−1	−1	−2	+1/3	+1/3	+4/3	−2/3
Q	0	−1	−1	+2/3	−1/3	+2/3	−1/3

Due to the different multiplicities, the generators are different for the left-handed and right-handed representations,

$$I_a^L = \frac{1}{2}\sigma_a, \quad I_a^R = 0; \tag{6.46}$$

the hypercharge representations are determined by $Y = y\mathbb{1}$ with the corresponding eigenvalue y from Table 6.2. The gauge transformations of $SU(2) \times U(1)$ are represented accordingly by the unitary matrices

$$U_{L,R}(\alpha_a, \alpha_y) = e^{i\alpha_a I_a^{L,R}} e^{i\alpha_y Y} \quad \text{with } \alpha_1, \alpha_2, \alpha_3, \alpha_y \in \mathbb{R}. \tag{6.47}$$

An immediate consequence concerns the masses of fermions: a mass term in $\mathcal{L}_0$ of the type

$$m\overline{\psi}\psi = m\left(\overline{\psi}_L\psi_R + \overline{\psi}_R\psi_L\right) \tag{6.48}$$

always connects left- and right-handed representations of fermion fields and thus is not invariant under gauge transformations

$$\psi_L \to U_L\,\psi_L, \quad \psi_R \to U_R\,\psi_R \tag{6.49}$$

with $U_{L,R}$ according to Eq. (6.47). In order to realize the global symmetry in $\mathcal{L}_0$ all fermion masses have to be set to zero. They will be reintroduced later in a gauge invariant way by means of the concept of Yukawa interactions.

We are now able to specify a free Lagrangian for the fundamental fermions that is invariant under the global gauge transformations (6.49). All the following steps are executed using the example of the first lepton and quark family and are valid in the same way also for the next generations. For simplification the notation $\nu \equiv \nu_e$ for the neutrino field is introduced. Due to $m_f = 0$ the free Lagrangian $\mathcal{L}_0$ consists of kinetic terms only,

$$\mathcal{L}_0 = \left(\overline{\nu}_L, \overline{e}_L\right) i\gamma^\mu \partial_\mu \begin{pmatrix} \nu_L \\ e_L \end{pmatrix} + \overline{e}_R\, i\gamma^\mu \partial_\mu e_R$$

$$+ \left(\overline{u}_L, \overline{d}_L\right) i\gamma^\mu \partial_\mu \begin{pmatrix} u_L \\ d_L \end{pmatrix} + \overline{u}_R\, i\gamma^\mu \partial_\mu u_R + \overline{d}_R\, i\gamma^\mu \partial_\mu d_R \tag{6.50}$$

being obviously invariant under the transformations (6.49).

6.4.2 *Local gauge invariance*

The next step towards invariance under local gauge transformations is done by means of the minimal substitution $\partial_\mu \to D_\mu$ with suitable covariant derivatives in $\mathcal{L}_0$. For each generator a vector field as the corresponding gauge field is introduced,

$$I_a \leftrightarrow W_\mu^a \quad (a = 1, 2, 3)$$
$$Y \leftrightarrow B_\mu. \tag{6.51}$$

The unequal representations of left- and right-handed fermions imply that also the respective covariant derivatives are different,

$$D_\mu^L = \partial_\mu - ig_2 \, I_a^L \, W_\mu^a + ig_1 \, \frac{Y}{2} B_\mu,$$
$$D_\mu^R = \partial_\mu + ig_1 \, \frac{Y}{2} B_\mu. \tag{6.52}$$

Due to the direct product of two groups there are two independent gauge coupling constants: g_2 for $SU(2)$, and g_1 for $U(1)$. The different sign of g_2 and g_1 in Eq. (6.52) is by convention only, as well as the use of $\frac{Y}{2}$ instead of Y as the generator (motivated by the Gell-Mann–Nishijima relation).

The minimal substitution in Eq. (6.50),

$$\mathcal{L}_0 \rightarrow \left(\overline{\nu}_L, \overline{e}_L \right) i\gamma^\mu D_\mu^L \begin{pmatrix} \nu_L \\ e_L \end{pmatrix} + \overline{e}_R \, i\gamma^\mu D_\mu^R e_R$$

$$+ \left(\overline{u}_L, \overline{d}_L \right) i\gamma^\mu D_\mu^L \begin{pmatrix} u_L \\ d_L \end{pmatrix} + \overline{u}_R \, i\gamma^\mu D_\mu^R u_R + \overline{d}_R \, i\gamma^\mu D_\mu^R d_R$$

$$= \mathcal{L}_0 + \frac{g_2}{2} \left(\overline{\nu}_L, \overline{e}_L \right) \gamma^\mu \, \sigma_a \begin{pmatrix} \nu_L \\ e_L \end{pmatrix} W_\mu^a$$

$$+ \frac{g_1}{2} \left(\overline{\nu}_L, \overline{e}_L \right) \gamma^\mu \begin{pmatrix} \nu_L \\ e_L \end{pmatrix} B_\mu + g_1 \, \overline{e}_R \, \gamma^\mu \, e_R \, B_\mu$$

$$+ \frac{g_2}{2} \left(\overline{u}_L, \overline{d}_L \right) \gamma^\mu \, \sigma_a \begin{pmatrix} u_L \\ d_L \end{pmatrix} W_\mu^a$$

$$- \frac{g_1}{6} \left(\overline{u}_L, \overline{d}_L \right) \gamma^\mu \begin{pmatrix} u_L \\ d_L \end{pmatrix} B_\mu - \frac{2}{3} g_1 \, \overline{u}_R \, \gamma^\mu \, u_R \, B_\mu + \frac{g_1}{3} \, \overline{d}_R \, \gamma^\mu \, d_R \, B_\mu$$

$$= \mathcal{L}_0 + \mathcal{L}_{\text{int}} \tag{6.53}$$

induces an interaction term between fermions and gauge fields. After introducing the complex linear combinations

$$W_\mu^\pm = \frac{1}{\sqrt{2}} \left(W_\mu^1 \mp iW_\mu^2 \right) \tag{6.54}$$

and making use of the indentity

$$\frac{1}{2}\left(\sigma_1 W_\mu^1 + \sigma_2 W_\mu^2\right) = \frac{1}{\sqrt{2}}\left(I_+ W_\mu^+ + I_- W_\mu^-\right)$$

with the matrices (6.39), one obtains

$$\mathcal{L}_{\text{int}} = \frac{g_2}{\sqrt{2}}\left[\left(\overline{\nu}_L, \overline{e}_L\right)\gamma^\mu\left(I_+ W_\mu^+ + I_- W_\mu^-\right)\begin{pmatrix}\nu_L\\e_L\end{pmatrix}\right.$$

$$\left.+\left(\overline{u}_L, \overline{d}_L\right)\gamma^\mu\left(I_+ W_\mu^+ + I_- W_\mu^-\right)\begin{pmatrix}u_L\\d_L\end{pmatrix}\right]$$

$$+\left(\overline{\nu}_L\gamma^\mu\nu_L\right)\cdot\frac{1}{2}\left(g_2 W_\mu^3 + g_1 B_\mu\right)$$

$$-\left(\overline{e}_L\gamma^\mu e_L\right)\cdot\frac{1}{2}\left(g_2 W_\mu^3 - g_1 B_\mu\right) + \left(\overline{e}_R\gamma^\mu e_R\right)\cdot g_1 B_\mu$$

$$+\left(\overline{u}_L\gamma^\mu u_L\right)\cdot\frac{1}{2}\left(g_2 W_\mu^3 - \frac{g_1}{3}B_\mu\right) - \frac{2}{3}\left(\overline{u}_R\gamma^\mu u_R\right)\cdot g_1 B_\mu$$

$$-\left(\overline{d}_L\gamma^\mu d_L\right)\cdot\frac{1}{2}\left(g_2 W_\mu^3 + \frac{g_1}{3}B_\mu\right) + \frac{1}{3}\left(\overline{d}_R\gamma^\mu d_R\right)\cdot g_1 B_\mu. \qquad (6.55)$$

The first two lines recover the charged-current interaction (6.25) and (6.31) when g_2 is identified with the coupling constant g of the V–A theory; the fields $W_\mu^\pm$ describe charged gauge bosons. The subsequent lines correspond to interactions with the neutral gauge fields W_μ^3 and B_μ. None of them can be assigned to the photon since the photon neither couples to neutrinos nor distinguishes between left- and right-handed fermions. Nevertheless, the electromagnetic interaction should somehow be contained in the neutral sector of $\mathcal{L}_{\text{int}}$.

Since the charge Q is a combination of I_3 and Y, an ansatz for the photon field as a linear combination of W_μ^3 and B_μ is evident. An orthogonal transformation in terms of a rotation yields two vector fields A_μ and Z_μ,

$$\begin{pmatrix}B_\mu\\W_\mu^3\end{pmatrix} = \begin{pmatrix}\cos\theta_W & \sin\theta_W\\-\sin\theta_W & \cos\theta_W\end{pmatrix}\begin{pmatrix}A_\mu\\Z_\mu\end{pmatrix}, \qquad (6.56)$$

where one of them shall be the electromagnetic field; the other orthogonal combination Z_μ then has to be assigned to the weak interaction as a neutral vector field. The rotation angle θ_W is denoted as the *electroweak mixing angle* or *weak mixing angle*. Convenient abbreviations are the notations

$$s_W = \sin\theta_W, \quad c_W = \cos\theta_W. \tag{6.57}$$

Inserting the rotation (6.56) into $\mathcal{L}_{\text{int}}$ yields for the neutrino part of the neutral sector in Eq. (6.55) the following expression,

$$\mathcal{L}_{\text{int},\nu} = \frac{1}{2}\left(\bar{\nu}_L\gamma^\mu\nu_L\right)\cdot\left[(-s_W g_2 + c_W g_1)\,A_\mu + (c_W g_2 + s_W g_1)\,Z_\mu\right].$$

If A_μ is the photon field, then the coefficient of A_μ is the coupling to the neutrino and has to be zero. This implies a correlation between θ_W and the gauge coupling constants,

$$s_W g_2 = c_W g_1. \tag{6.58}$$

Hence, for the neutral neutrino interaction only the coupling to the Z_μ remains, according to

$$\frac{g_2}{2c_W}\left(\bar{\nu}_L\gamma^\mu\nu_L\right)Z_\mu = \frac{g_2}{2c_W}\left(\bar{\nu}\,\gamma^\mu\frac{1-\gamma_5}{2}\,\nu\right)Z_\mu. \tag{6.59}$$

Concerning the electron part in the neutral sector of $\mathcal{L}_{\text{int}}$, the coupling to photon and Z can be written as follows, using Eq. (6.58),

$$\mathcal{L}_{\text{int},e} = (\bar{e}_L\gamma^\mu e_L)\cdot\left[s_W g_2\,A_\mu + \frac{2s_W^2 - 1}{2c_W}g_2\,Z_\mu\right]$$

$$+ (\bar{e}_R\gamma^\mu e_R)\cdot\left[s_W g_2\,A_\mu + \frac{s_W^2}{c_W}g_2\,Z_\mu\right]$$

$$= (\bar{e}_L\gamma^\mu e_L + \bar{e}_R\gamma^\mu e_R)\,(s_W g_2)\,A_\mu$$

$$+ \frac{g_2}{2c_W}\left[(\bar{e}_L\gamma^\mu e_L)\,(2s_W^2 - 1) + (\bar{e}_R\gamma^\mu e_R)\,(2s_W^2)\right]Z_\mu$$

$$= (\bar{e}\gamma^\mu e)\,(s_W g_2)\,A_\mu + \frac{g_2}{4c_W}\left[(\bar{e}\gamma^\mu e)\,(4s_W^2 - 1) + (\bar{e}\gamma^\mu\gamma_5 e)\right]Z_\mu.$$

The A_μ field has a left-right symmetric coupling to the charged fermions and hence can be interpreted as the photon field once the combination $s_W g_2$ is identified with the elementary charge e. In this way one recovers the electromagnetic interaction as a part of the gauge interaction based on $SU(2)\times U(1)$ and, on top, one finds the essential correlation

$$\boxed{g_2 = \frac{e}{\sin\theta_W}} \tag{6.60}$$

connecting the $SU(2)$ coupling constant of the weak interaction with the electromagnetic coupling constant via the electroweak mixing angle.

The other part of the neutral gauge interaction, the coupling to the Z_μ field, is different for left-handed and right-handed currents, though not of the type V–A. Often vector and axialvector currents instead of the chiral currents are used, with coupling constants v_e, a_e defined according to the normalization

$$\frac{g_2}{4c_W} \left[(\bar{e}\gamma^\mu e)\, (4s_W^2 - 1) + (\bar{e}\gamma^\mu\gamma_5 e) \right] Z_\mu = \frac{g_2}{2c_W} \left[(\bar{e}\gamma^\mu e)\, v_e - (\bar{e}\gamma^\mu\gamma_5 e)\, a_e \right] Z_\mu$$

$$= \frac{g_2}{2c_W} \left[\bar{e}\,\gamma^\mu (v_e - a_e\gamma_5)\, e \right] Z_\mu$$

$$(6.61)$$

and expressed as follows

$$v_e = -\frac{1}{2} + 2s_W^2 = I_3^e - 2Q_e s_W^2,$$

$$a_e = -\frac{1}{2} = I_3^e$$

$$(6.62)$$

in terms of the quantum numbers, charge $Q_e = -1$ and third component of the isospin $I_3^e = -\frac{1}{2}$, of the electron field, as specified in Table 6.2.

Performing the same steps for the quarks, starting from Eq. (6.55) and with some elementary reshufflings, one obtains the quark couplings to the electromagnetic field A_μ and to the neutral weak field Z_μ in a form analogous to the leptons. The currents that couple to the Z field do not change the charge of the involved fermions; in the terminology of the weak interaction they are thus denoted as *neutral currents* (NC), supplementing the charged currents that were known empirically already before the Standard Model. The Lagrangian (6.55) of the interaction is separated accordingly into three parts,

$$\mathcal{L}_{\text{int}} = \mathcal{L}_{\text{int}}^{(CC)} + \mathcal{L}_{\text{int}}^{(NC)} + \mathcal{L}_{\text{int}}^{(em)},$$

where the CC part is given by the already known first two lines of Eq. (6.55). The other terms describe the electromagnetic and the NC interaction,

$$\mathcal{L}_{\text{int}}^{(em)} = -e \sum_f Q_f \left(\overline{\psi}_f\, \gamma^\mu\, \psi_f \right) A^\mu, \quad f = e, u, d$$

$$\mathcal{L}_{\text{int}}^{(NC)} = \frac{g_2}{2c_W} \sum_f \left[\overline{\psi}_f\, \gamma^\mu\, (v_f - a_f\gamma_5)\, \psi_f \right] Z_\mu, \quad f = \nu, e, u, d$$

$$(6.63)$$

with the vector and axialvector couplings in generalization of Eq. (6.62),

$$\boxed{v_f = I_3^f - 2Q_f s_W^2, \quad a_f = I_3^f}. \tag{6.64}$$

Although the interaction terms given above refer to the first generation of leptons and quarks, the summation can be immediately extended to cover also the other two generations, with the same couplings. For the CC interaction of quarks the augmentation to three generations shows a peculiarity that has its origin in the mass terms. It will be treated later in Sec. 6.4.5.

Feynman rules. The interaction terms can be translated into vertices for the Feynman rules. The CC vertices are the same as in the $V-A$ theory, with $g = g_2$, and will not be listed here again. The special feature mentioned above for quarks with three generations consists in supplementing the CC vertices by the elements of the CKM mixing matrix and will be done later.

For the interaction of fermions with photon and Z boson the vertices are given by (for all fermions f)

$$-i\,e\,Q_f\,\gamma^\mu$$

$$i\,\frac{g_2}{2c_W}\,\gamma^\mu\left(v_f - a_f\gamma_5\right)$$

Remark. The QED vertex has a different sign as in the QED Feynman rules in Sec. 3.5. This is a consequence of the sign convention in the covariant derivative; it corresponds to a global sign and is without physical relevance.

6.4.3 *Dynamics of the gauge fields*

For specifying the intrinsic dynamics of the gauge bosons one starts from the original fields W_μ^a and B_μ. For each multiplet there is a field strength tensor

$$W_{\mu\nu}^a = \partial_\mu W_\nu^a - \partial_\nu W_\mu^a + g_2 \, \epsilon_{abc} \, W_\mu^b \, W_\nu^c,$$

$$B_{\mu\nu} = \partial_\mu B_\nu - \partial_\nu B_\mu,$$

(6.65)

and the corresponding Lagrangian is, altogether, given by

$$\mathcal{L}_{W,B} = -\frac{1}{4} \sum_a W_{\mu\nu}^a \, W^{a,\mu\nu} - \frac{1}{4} B_{\mu\nu} B^{\mu\nu}. \tag{6.66}$$

The quadratic parts determine the propagators of the free fields; the non-Abelian cubic and quartic terms describe the self interactions of the W fields related to the isospin. Performing the transformation $W_\mu^a, B_\mu \to W_\mu^\pm, Z_\mu, A_\mu$ according to Eqs. (6.54) and (6.56) and introducing the notation

$$F_{\mu\nu} = \partial_\mu A_\nu - \partial_\nu A_\mu, \quad Z_{\mu\nu} = \partial_\mu Z_\nu - \partial_\nu Z_\mu$$

one obtains the trilinear and quadrilinear self couplings expressed in terms of the physical vector fields ($h.c. = hermitean\ conjugate$),

$$
\begin{aligned}
\mathcal{L}_{\mathrm{V,self}} = {} & e\left[(\partial_\mu W_\nu^+ - \partial_\nu W_\mu^+)\, W^{-\mu} A^\nu + W_\mu^+ W_\nu^- \, F^{\mu\nu} + h.c.\right] \\
& + g_2\, c_W\left[(\partial_\mu W_\nu^+ - \partial_\nu W_\mu^+)\, W^{-\mu} Z^\nu + W_\mu^+ W_\nu^- \, Z^{\mu\nu} + h.c.\right] \\
& - \frac{g_2^2}{4}\left[(W_\mu^- W_\nu^+ - W_\nu^- W_\mu^+)W_\mu^+ W_\nu^- + h.c.\right] \\
& - \frac{e^2}{4}\left(W_\mu^+ A_\nu - W_\nu^+ A_\mu\right)\left(W^{-\mu} A^\nu - W^{-\nu} A^\mu\right) \\
& - \frac{g_2^2}{4} c_W^2 \left(W_\mu^+ Z_\nu - W_\nu^+ Z_\mu\right)\left(W^{-\mu} Z^\nu - W^{-\nu} Z^\mu\right) \\
& + \frac{e g_2}{2} c_W \left(W_\mu^+ A_\nu - W_\nu^+ A_\mu\right)\left(W^{-\mu} Z^\nu - W^{-\nu} Z^\mu\right) + h.c.
\end{aligned}
$$

(6.67)

yielding the following vertices for the electroweak gauge self interactions (all momenta and charges have to be understood as incoming).

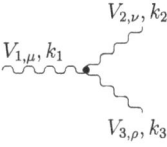

$$V_{2,\nu}, k_2 \qquad i\, C_{V_1 V_2 V_3}\left[g_{\mu\nu}(k_2-k_1)_\rho + g_{\nu\rho}(k_3-k_2)_\mu + g_{\rho\mu}(k_1-k_3)_\nu\right]$$

$V_{1,\mu}, k_1$

$V_{3,\rho}, k_3$

$V_1 V_2 V_3$	$C_{V_1 V_2 V_3}$
AW^+W^-	$-e$
ZW^+W^-	$g_2\, c_W$

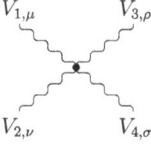

$V_{1,\mu}$ $\qquad$ $V_{3,\rho}$ $\qquad$ $i\, C_{V_1 V_2 V_3 V_4}\left(2g_{\mu\nu}g_{\rho\sigma} - g_{\mu\rho}g_{\nu\sigma} - g_{\nu\rho}g_{\mu\sigma}\right)$

$V_{2,\nu}$ $\qquad$ $V_{4,\sigma}$

$V_1 V_2 V_3 V_4$	$C_{V_1 V_2 V_3 V_4}$
$W^+W^+W^-W^-$	g_2^2
W^+W^-ZZ	$-g_2^2\, c_W^2$
W^+W^-AZ	$eg_2\, c_W$
W^+W^-AA	$-e^2$

The coefficients are exclusively determined by gauge symmetry. Deviations from these values can only be of non-standard origin, such as remnants from new physics at some higher mass scale.

6.4.4 Gauge-boson masses and Higgs mechanism

At our present level of the description of the electroweak interaction the gauge bosons $\gamma, Z, W^\pm$ are massless. In physics reality, however, this applies only to the photon, whereas the $W^\pm$ and Z bosons are massive; this has been empirically known since 1983 and was deduced even earlier from the low-energy behaviour of the weak interaction. The naive procedure to overcome the problem, just adding mass terms like $M_W^2\, W_\mu^+ W^{-\,\mu}$ and $M_Z^2\, Z_\mu Z^\mu$ to the gauge-symmetric Lagrangian, is not an acceptable concept.

- As explicated in Sec. 6.3.5, the longitudinal polarization vectors of massive vector bosons at high energies become proportional to their momenta, $\epsilon_L^\mu = k^\mu/M_V$. This causes an unlimited increase of the cross sections for scattering of longitudinally polarized vector bosons, $V_L V_L \to V_L V_L$, with energy, in clear contradiction to the unitarity of the S-matrix.
- In the calculation of Feynman diagrams with loops in higher order of perturbation theory, diagrams with massive vector bosons within the loops generate uncontrollable divergences from integration over the loop momentum (ultraviolet divergences), owing to the momentum-dependent

term in the numerator of the propagator in Eq. (6.26). Such divergences require additional ad-hoc cutoff parameters and thus destroy the predictive power in higher orders of perturbation theory, which for massless gauge bosons is ensured (denoted as *renormalizability*).

The formal reason for the problems related to mass $\neq 0$ is the breaking of local gauge symmetry by the mass terms. Local gauge invariance of the Lagrangian $\mathcal{L}$ is essential for the step from the classical theory to a quantum field theory (see Sec. 4.4.4). The augmentation of $\mathcal{L}$ by introducing a gauge-fixing term and the associated ghost field Lagrangian extend the original local gauge symmetry to the BRST symmetry of the entire Lagrangian which implies relations between Green functions[1] independent of the order of perturbation theory. These *Slavnov-Taylor identities* are of basic importance for the internal consistency of the theory; they ensure renormalizability and guarantee unitarity and gauge invariance of S-matrix elements to all orders.

A viable concept for the introduction of mass terms for the gauge bosons hence has to be compatible with the local gauge invariance of the Lagrangian $\mathcal{L}$. The construction applied in the Standard Model, known by the keyword *Higgs mechanism*, is based on the extension of the current system of fermions and gauge bosons by an extra scalar field with a gauge invariant Lagrangian, however, with a ground state not symmetric under gauge transformations. This kind of breaking the original symmetry is the concept of *spontaneous symmetry breaking*. It will be elucidated first in terms of a simplified model [Higgs (1964); Guralnik (1964); Englert (1964)].

6.4.4.1 *Abelian model*

Consider the example of a spontaeously broken $U(1)$ symmetry by means of a complex scalar field $\phi(x)$, described by the Lagrangian

$$\mathcal{L} = \left(\partial_\mu \phi\right)^\dagger \left(\partial^\mu \phi\right) - V\left(\phi\right). \tag{6.68}$$

The field has a self-interaction in terms of the potential V in Fig. 6.1, with a minimum at $\phi_0 = v$ corresponding to the *vacuum expectation value* of the field in the ground state, $\langle 0|\phi_0|0\rangle$. Note: The field configuration $\phi(x) \equiv \phi_0$

---

[1]Green functions or n-point functions $\langle 0|T\phi_1(x_1)\,\phi_2(x_2)\cdots\phi_n(x_n)|0\rangle$ are generalizations of the 2-point functions like (4.129) to n fields.

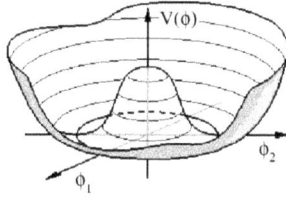

Fig. 6.1 Scalar potential $V(\phi)$, with $\phi_1 = \mathrm{Re}\,\phi$, $\phi_2 = \mathrm{Im}\,\phi$.

determines the ground state since the Hamiltonian density

$$\mathcal{H} = |\partial_0\phi|^2 + |\nabla\phi|^2 + V(\phi) \tag{6.69}$$

adopts its minimum at $\phi = \phi_0$.

For $V = V(|\phi|)$ the Lagrangian $\mathcal{L}$ is symmetric under $U(1)$ transformations,

$$\phi \to \phi' = e^{i\alpha}\,\phi \quad \Rightarrow \quad \mathcal{L} \to \mathcal{L}' = \mathcal{L}. \tag{6.70}$$

On the other hand, in case of $v \neq 0$ the value of the field at the minimum is not symmetric,

$$\phi_0 \to \phi_0' = e^{i\alpha}\,v \neq \phi_0, \tag{6.71}$$

i.e. the $U(1)$ symmetry is broken spontaneously. Furthermore, one has the relation

$$V(\phi_0') = V(\phi_0), \tag{6.72}$$

i.e. the vacuum is degenerate. For the next steps the field ϕ is expressed in terms of its absolute value and phase,

$$\phi(x) = \eta(x)e^{i\theta(x)}, \quad \eta(x), \theta(x) \in \mathbb{R}$$

and η is expanded around v,

$$\eta(x) = v + \frac{1}{\sqrt{2}}\,H(x). \tag{6.73}$$

The potential $V(\phi) = V(\eta)$ at the minimum obeys $\quad V'(v) = 0,\ V''(v) > 0$, hence V can be expanded around the minimum according to

$$V(\eta) = V(v) + \frac{1}{2}V''(v)\cdot\frac{1}{2}H^2 + \cdots$$

Hence, one obtains for the quadratic part of $\mathcal{L}$ the expression

$$\mathcal{L} = \frac{1}{2}(\partial_\mu H)(\partial^\mu H) - \underbrace{\frac{1}{2}V''(v)}_{= \; m_H^2} \cdot \frac{1}{2}H^2 + v^2(\partial_\mu \theta)(\partial^\mu \theta) + \cdots$$

giving evidence to two notable statements.

(i) The real H-field is massive; when quantized, it describes neutral particles with spin 0 and mass m_H.
(ii) The θ-field is massless since there is no θ^2 term in $\mathcal{L}$. This field is also named a *Goldstone field*, the associated massless spin-0 particles are denoted as *Goldstone bosons*.

This issue can be generalized to systems where a global symmetry with several generators $T_1, \ldots T_N$ is spontaneously broken by the ground state with vacuum expectation value ϕ_0. The following statement holds:

The partial symmetry corresponding to a generator T_k is spontaneously broken if and only if $T_k \phi_0 \neq 0$. This is obvious because of

$$\phi_0' = e^{i\alpha_k T_k}\phi_0 \neq \phi_0 \iff T_k \phi_0 \neq 0.$$

Generators T_m of an unbroken (partial) symmetry always obey $T_m \phi_0 = 0$.

Statement (ii) given above turns out to be a special case of a general theorem [Goldstone (1961, 1962)] concerning spontaneously broken global symmetries, known as the *Goldstone Theorem*:

For each broken generator T_k of a global symmetry there is a massless scalar Goldstone field $\theta_k(x)$.

Of special interest is the situation when the broken symmetry is a local gauge symmetry. To this end we consider an extension of the Abelian model described by Eq. (6.68) to a model involving an additional vector field A_μ, with the Lagrangian

$$\mathcal{L} = (D_\mu\phi)^\dagger(D^\mu\phi) - V(\phi) - \frac{1}{4}F_{\mu\nu}F^{\mu\nu} \quad \text{with } D_\mu = \partial_\mu - ieA_\mu, \quad (6.74)$$

invariant under local $U(1)$ transformations

$$\phi'(x) = e^{i\alpha(x)}\,\phi(x) = e^{i\alpha(x)}\,e^{i\theta(x)}\eta(x)$$

$$A'_\mu(x) = A_\mu(x) + \frac{1}{e}\partial_\mu\alpha(x)$$

with an arbitrary real function $\alpha(x)$. By the specific choice $\alpha(x) = -\theta(x)$ one obtains $\phi'(x) = \eta(x)$ and

$$\mathcal{L} = \left[(\partial_\mu + ieA'_\mu)\eta\right]\left[(\partial^\mu - ieA'^\mu)\eta\right] - V(\eta) - \frac{1}{4}F'_{\mu\nu}F'^{\mu\nu}$$

without the field $\theta(x)$, which is unphysical since it has been eliminated by a gauge transformation. Inserting the expansion (6.73) of the η field yields for the quadratic part of $\mathcal{L}$ the expression

$$\mathcal{L} = \left|(\partial_\mu - ieA'_\mu)(v + \frac{1}{\sqrt{2}}H)\right|^2 - \frac{1}{4}F'_{\mu\nu}F'^{\mu\nu} - V$$

$$= \underbrace{-\frac{1}{4}F'_{\mu\nu}F'^{\mu\nu} + v^2e^2\,A'_\mu A'^\mu}_{\text{massive } A \text{ field}} + \frac{1}{2}(\partial_\mu H)(\partial^\mu H) - \frac{m_H^2}{2}H^2 + \cdots$$

from which one can see that the gauge field A_μ has got a mass term now, with a mass $m_A \sim ev$ generated by the coupling of the gauge field to the vacuum of the scalar field. Thus one obtains a massive vector field without spoiling the local gauge invariance of $\mathcal{L}$. The specific gauge chosen above is the *unitary gauge* where the Goldstone field has been eliminated; it contains exclusively the physical degrees of freedom of the system. In the unitary gauge the propagator of the vector field is given by the expression (6.26), with the replacement $M_W \to m_A$.

The massive vector field now has got a longitudinal polarization state (refering to helicity $= 0$) as well, and the polarization sum (6.34) (with $M_W \to m_A$) is the projector on the subspace orthogonal to the momentum k. In comparison to a massless gauge field with two transverse polarization states only, the massive field has one more degree of freedom; in exchange, the scalar field has one degree of freedom less due to the absent Goldstone field, and thus the number of degrees of freedom in total is the same. The Goldstone boson has been converted to become the helicity state with $\lambda = 0$ of the gauge boson.

6.4.4.2 $SU(2) \times U(1)$ model

For spontaneous breaking of the local gauge symmetry in the Standard Model one has to take into account that the electromagnetic partial symmetry $U(1)_{em}$ with the generator Q remains unbroken in order to keep the photon massless. Since for each of $W^\pm, Z$ a longitudinal polarization state will exist when they acquire masses, three Goldstone fields are required. The minimal possibility to implement such a configuration needs an isospin doublet of complex scalar fields

$$\Phi(x) = \begin{pmatrix} \phi^+(x) \\ \phi^0(x) \end{pmatrix} \tag{6.75}$$

with the quantum numbers $I = \frac{1}{2}$, $Y = 1$. The charge Q as the generator of $U(1)_{em}$ is given via the Gellmann–Nishijima relation as follows,

$$Q = I_3 + \frac{Y}{2} = \begin{pmatrix} 1 & 0 \\ 0 & 0 \end{pmatrix}. \tag{6.76}$$

ϕ^+ is a charged field with electromagnetic charge $+1$ and ϕ^0 is a neutral field, since

$$Q \begin{pmatrix} \phi^+ \\ 0 \end{pmatrix} = \begin{pmatrix} \phi^+ \\ 0 \end{pmatrix}, \quad Q \begin{pmatrix} 0 \\ \phi^0 \end{pmatrix} = 0.$$

For spontaneous symmetry breaking the existence of a non-zero vacuum expectation value Φ_0 of the scalar doublet is needed. The unbroken $U(1)_{em}$ symmetry requires

$$Q\, \Phi_0 = 0$$

which implies that the charged component of Φ_0 has to vanish (the normalization factor is by convention),

$$\Phi_0 = \frac{1}{\sqrt{2}} \begin{pmatrix} 0 \\ v \end{pmatrix}. \tag{6.77}$$

The potential for describing the self-interaction of the Φ field is chosen as a polynomial that is invariant under the entire gauge group, globally and locally,

$$V = -\mu^2 \left(\Phi^\dagger \Phi\right) + \frac{\lambda}{4} \left(\Phi^\dagger \Phi\right)^2 \tag{6.78}$$

with real parameters $\mu > 0$, $\lambda > 0$ and

$$\Phi^\dagger = \left(\phi^-, \phi^{0\,\dagger}\right), \quad \text{where } \phi^- = (\phi^+)^\dagger.$$

V has a minimum at the value $\Phi_0 \neq 0$, which is determined by v within the ansatz (6.77). Expressed in terms of the parameters of the potential, v is given by

$$v = \frac{2\mu}{\sqrt{\lambda}}. \tag{6.79}$$

The choice of the potential as the polynomial (6.78) has a pragmatic reason. It is the minimal version leading to spontaneous symmetry breaking and obeying the following conditions,

- V is restricted fom below, hence there is a state of lowest energy of the system;
- the interaction terms in V are renormalizable, i.e. they enable perturbative calculations at higher orders without additional cutoff parameters.

The gauge invariant coupling of Φ to the gauge fields follows the rule of introducing a covariant derivative determined by isospin and hypercharge of Φ according to

$$D_\mu = \partial_\mu - ig_2 \frac{\sigma_a}{2} W_\mu^a + i\frac{g_1}{2} B_\mu,$$

yielding the Lagrangian

$$\mathcal{L}_H = \left(D_\mu \Phi\right)^\dagger \left(D_\mu \Phi\right) - V. \tag{6.80}$$

Exploiting the local gauge invariance of $\mathcal{L}$, a particularly simple expression for Φ can be found. To this end the doublet (6.75), after separating off the vacuum Φ_0, is decomposed into real fields ϕ_1, ϕ_2, χ, H in the following way,

$$\Phi(x) = \begin{pmatrix} \phi^+(x) \\ \frac{1}{\sqrt{2}}\left(v + H(x) + i\chi(x)\right) \end{pmatrix} = \begin{pmatrix} \phi_1(x) + i\,\phi_2(x) \\ \frac{1}{\sqrt{2}}\left(v + H(x) + i\chi(x)\right) \end{pmatrix}. \tag{6.81}$$

One finds that by a gauge transformation with the local parameters ϕ_1, ϕ_2, χ the following form can be achieved,

$$\Phi(x) \quad \rightarrow \quad U\left(\phi_1, \phi_2, \chi\right)\Phi(x) = \begin{pmatrix} 0 \\ \frac{1}{\sqrt{2}}\left(v + H(x)\right) \end{pmatrix} \tag{6.82}$$

corresponding to the *unitary gauge*. The components χ as well as ϕ_1, ϕ_2 or ϕ^+, ϕ^-, respectively, have been eliminated by the gauge transformation and thus prove to be the unphysical Goldstone fields. As such they have been converted into the longitudinal polarization states of Z, W^+, W^-. The field H is physical, with a mass term corresponding to $M_H = \mu\sqrt{2}$. Hence, H describes neutral spinless particles with mass M_H. In the usual terminology H is the *Higgs field*, the associated particles are called *Higgs bosons*.

For the further discussion it is sufficient to remain in the unitary gauge. With the corresponding covariant derivative

$$D_\mu \Phi = \frac{1}{\sqrt{2}} \begin{pmatrix} 0 \\ \partial_\mu H \end{pmatrix} + i\frac{g_2}{2} W_\mu^+ \begin{pmatrix} v + H \\ 0 \end{pmatrix} + \frac{i}{2\sqrt{2}} (g_2 W_\mu^3 + g_1 B_\mu) \begin{pmatrix} 0 \\ v + H \end{pmatrix},$$

the kinetic part of $\mathcal{L}_H$ turns into

$$(D_\mu \Phi)^\dagger (D_\mu \Phi) = \frac{1}{2}(\partial_\mu H)(\partial^\mu H) + \frac{g_2^2}{4}(v+H)^2 W_\mu^- W^{+\mu}$$

$$+ \frac{1}{8}(v+H)^2 (B_\mu, W_\mu^3) \begin{pmatrix} g_1^2 & g_1 g_2 \\ g_1 g_2 & g_2^2 \end{pmatrix} \begin{pmatrix} B^\mu \\ W^{3\mu} \end{pmatrix}. \quad (6.83)$$

The rotation (6.56) with the mixing angle θ_W introduces the physical fields A_μ, Z_μ and diagonalizes the matrix above by means of the relation (6.58), yielding

$$(B_\mu, W_\mu^3) \begin{pmatrix} g_1^2 & g_1 g_2 \\ g_1 g_2 & g_2^2 \end{pmatrix} \begin{pmatrix} B^\mu \\ W^{3\mu} \end{pmatrix} = (A_\mu, Z_\mu) \begin{pmatrix} 0 & 0 \\ 0 & g_1^2 + g_2^2 \end{pmatrix} \begin{pmatrix} A^\mu \\ Z^\mu \end{pmatrix}.$$

After diagonalization, the quadratic part of $\mathcal{L}_H$ containing v^2,

$$(D_\mu \Phi)^\dagger (D_\mu \Phi)\Big|_{v^2} = \frac{g_2^2}{4} v^2 W_\mu^- W^{+\mu} + \frac{1}{2} \cdot \frac{v^2}{4} (g_1^2 + g_2^2) Z_\mu Z^\mu$$

$$\equiv M_W^2 W_\mu^- W^{+\mu} + \frac{1}{2} M_Z^2 Z_\mu Z^\mu, \quad (6.84)$$

turns out to be a mass term for the $W^\pm$ and Z fields; the photon field remains massless, as desired. The masses are determined by

$$M_W = \frac{v}{2} g_2, \quad M_Z = \frac{v}{2}\sqrt{g_1^2 + g_2^2} \quad (6.85)$$

and establish a connection with the electroweak mixing angle,

$$\frac{M_W^2}{M_Z^2} = \frac{g_2^2}{g_2^2 + g_1^2} = \frac{1}{1 + \tan^2 \theta_W} = \cos^2 \theta_W$$

yielding the important correlation

$$\boxed{\cos \theta_W = \frac{M_W}{M_Z}}. \quad (6.86)$$

Furthermore, the kinetic scalar Lagrangian (6.83) contains interaction terms of the Higgs field with the gauge fields,

$$\mathcal{L}_{H,\text{int}} = \frac{g_2^2}{4}\left(2vH + H^2\right) W_\mu^- W^{+\mu} + \frac{g_1^2 + g_2^2}{8}\left(2vH + H^2\right) Z_\mu Z^\mu$$

$$= g_2 M_W \, H \, W_\mu^- W^{+\mu} + \frac{g_2^2}{4}\, H^2 \, W_\mu^- W^{+\mu}$$

$$+ \frac{g_2}{2c_W} M_Z \, H \, Z_\mu Z^\mu + \frac{g_2^2}{8c_W^2}\, H^2 \, Z_\mu Z^\mu \tag{6.87}$$

yielding additional vertices for the list of Feynman rules,

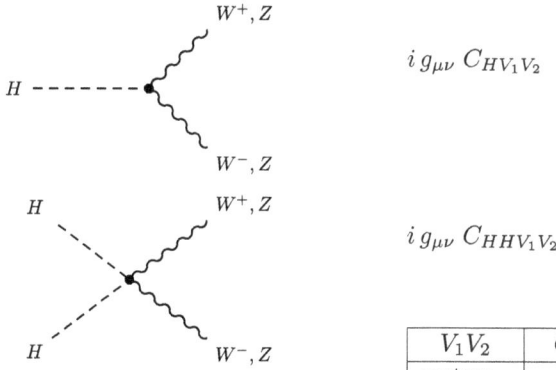

$$i\, g_{\mu\nu}\, C_{HV_1 V_2}$$

$$i\, g_{\mu\nu}\, C_{HHV_1 V_2}$$

$V_1 V_2$	$C_{HV_1 V_2}$	$C_{HHV_1 V_2}$
$W^+ W^-$	$g_2 M_W$	$g_2^2/2$
ZZ	$g_2 M_Z/c_W$	$g_2^2/(2c_W)$

Dynamics of Higgs and gauge fields. $\mathcal{L}_{W,B}$ and $\mathcal{L}_H$ determine the dynamics of the gauge fields and the Higgs field in close mutual connection, concerning the free dynamics as well as the interaction terms. The Lagrangian combined from the parts in Eqs. (6.66) and (6.80),

$$\mathcal{L}_{W,B} + \mathcal{L}_H = -\frac{1}{2}\left(\partial_\mu W_\nu^- - \partial_\nu W_\mu^-\right)\left(\partial^\mu W^{+\nu} - \partial^\nu W^{+\mu}\right) + M_W^2 \, W_\mu^- W^{+\mu}$$

$$-\frac{1}{4}\left(\partial_\mu Z_\nu - \partial_\nu Z_\mu\right)\left(\partial^\mu Z^\nu - \partial^\nu Z^\mu\right) + \frac{M_Z^2}{2}\, Z_\mu Z^\mu$$

$$-\frac{1}{4}\left(\partial_\mu A_\nu - \partial_\nu A_\mu\right)\left(\partial^\mu A^\nu - \partial^\nu A^\mu\right)$$

$$+\frac{1}{2}\left[(\partial_\mu H)(\partial^\mu H) - M_H^2 H^2\right]$$

$$+\,(\text{interaction terms}) \tag{6.88}$$

comprises as interaction terms the gauge boson self-couplings (6.67), the gauge boson–Higgs couplings (6.87), and the scalar self-couplings from the potential to be discussed later in Sec. 6.6. The quadratic terms in Eq. (6.88) determine the propagators of the free fields: the photon propagator as in QED, the W and Z propagators as in Eq. (6.26) with the respective masses M_W and M_Z, and as a new element the scalar propagator of the Higgs field. The propagator of the neutral Higgs field, in the Feynman rules displayed by the graphical symbol

$$i\,D(q) \qquad\qquad \text{-- -- -- -- -- --}$$
$$q$$

can be obtained as the Green function of the Klein–Gordon equation with causal boundary conditions, corresponding to the scalar 2-point function

$$\langle 0|T H(x)\,H(y)|0\rangle = i\,D(x-y).$$

In momentum space it is given by

$$i\,D(q) = \frac{i}{q^2 - M_H^2 + i\varepsilon}. \qquad (6.89)$$

The propagators and vertices listed so far have been obtained in the unitary gauge, which involves exclusively physical degrees of freedom and hence is rather compact and transparent. For higher-order calculations, but also for formal considerations, however, it is of advantage to apply a more general gauge where the Goldstone bosons are not eliminated. In such a gauge it is necessary to introduce a gauge-fixing term in the Lagrangian, as discussed in Sec. 4.4.4, otherwise the system of coupled equations for the propagators of the gauge and Goldstone fields has no solution. Frequently the class of R_ξ gauges is used extending the expression in Sec. 4.4.4 to the term

$$\mathcal{L}_{\text{fix}} = -F^- F^+ - \frac{1}{2}F_Z^2 - \frac{1}{2}F_A^2 \qquad (6.90)$$

with the following linear combinations of gauge fields and Goldstone fields,

$$F_A = \frac{1}{\sqrt{\xi_A}}\,\partial^\mu A_\mu, \quad F_Z = \frac{1}{\sqrt{\xi_Z}}\left(\partial^\mu Z_\mu - M_Z \xi_Z\,\chi\right),$$

$$F^\pm = \frac{1}{\sqrt{\xi_W}}\left(\partial^\mu W_\mu^\pm \mp i M_W \xi_W\,\phi^\pm\right),$$

containing ξ_W, ξ_Z, ξ_A as arbitrary real parameters. In the R_ξ gauge the vector boson propagators read (for $V = W, Z, A$ with $M_A = 0$)

$$D_{\mu\nu}^V(q) = \frac{i}{q^2 - M_V^2 + i\varepsilon} \left(-g_{\mu\nu} + (1 - \xi_V) \frac{q_\mu q_\nu}{q^2 - \xi_V M_V^2} \right)$$

$$= \frac{i}{q^2 - M_V^2 + i\varepsilon} \left(-g_{\mu\nu} + \frac{q_\mu q_\nu}{q^2} \right) - \frac{i\xi_V}{q^2 - \xi_V M_V^2} \frac{q_\mu q_\nu}{q^2}, \qquad (6.91)$$

without an increase by extra powers of the momentum, in contrast to the unitary gauge. They become particularly simple for $\xi_V = 1$, labeled as *'t Hooft–Feynman gauge*. In addition, there are propagators for the Goldstone fields, given by Eq. (6.89) with $M_H^2 \to \xi_W M_W^2$ for $\phi^\pm$ and $M_H^2 \to \xi_Z M_Z^2$ for χ. By the gauge condition the Goldstone bosons acquire a fictitious mass which for $\xi_V = 1$ coincides with the respective gauge boson mass. In the limit $\xi_V \to \infty$ one recovers the unitary gauge. The advantage of the apparently more complicated gauge fixing (6.90) is the cancellation of mixing terms between gauge fields and Goldstone fields originating from the kinetic part of the Higgs field Lagrangian $\mathcal{L}_H$ in Eq. (6.80).

Besides the enlarged set of propagators in the R_ξ gauge there are quite a few additional vertices involving the Goldstone bosons, originating from Eq. (6.80) as well. On top, for the cancellation of the unphysical gauge degrees of freedom a ghost contribution $\mathcal{L}_{\text{ghost}}$ to the Lagrangian is necessary, with unphysical ghost fields $u_V, \bar{u}_V$ for each $V = W, Z, A$, which couple to the gauge fields and to all components of the Higgs doublet. In the 't Hooft–Feynman gauge, the ghost field propagators are of the scalar type (6.89), but with fictitious mass terms M_V^2 in the denominator. Altogether one obtains a long list of Feynman rules with a large number of propagators and vertices, and an explanation would be far beyond the scope of this introduction. A complete set of Feynman rules in the $\xi = 1$ gauge can be found in [Denner (1993)].

By adding $\mathcal{L}_{\text{fix}}$ to the Lagrangian the local gauge symmetry is broken. On the other hand, adding $\mathcal{L}_{\text{fix}} + \mathcal{L}_{\text{ghost}}$ establishes the invariance of the entire Lagrangian under BRST transformations. As already pointed out, this symmetry at the quantum level ensures that all unphysical degrees of freedom including the gauge parameters ξ_V at the end cancel in the calculation of S-matrix elements for physical processes with on-shell external particles.

6.4.5 *Fermion masses and Yukawa interaction*

The final step in the construction of the electroweak Standard Model consists in the solution of the problem of fermion masses. These had to be set to zero so far because finite masses connect fermions of left- and right-handed chirality and thus violate the basic gauge symmetry. Meanwile, with the presence of the Higgs field, a tool exists that allows to introduce fermion masses in a gauge invariant way by a new coupling of the fermion fields to the scalar doublet. This kind of scalar–fermion interaction is named *Yukawa interaction*, according to nuclear physics where Yukawa in 1935 had proposed a pion–nucleon interaction in order to explain the nuclear force.

6.4.5.1 *Leptons*

Choosing the example of the first generation, the Yukawa interaction is given by

$$\mathcal{L}_Y = -g_e \left[\overline{\psi}_L \, \Phi \, e_R + \overline{e}_R \, \Phi^\dagger \psi_L \right] \tag{6.92}$$

with the scalar doublet Φ in (6.75), the left-handed lepton doublet

$$\psi_L = \begin{pmatrix} \nu_L \\ e_L \end{pmatrix}, \quad \overline{\psi}_L = (\overline{\nu}_L, \overline{e}_L),$$

and a coupling constant g_e, the Yukawa coupling constant, being a further independent free parameter. The physics content becomes clear most suitable by choosing the unitary gauge (6.82) for Φ, yielding

$$\mathcal{L}_Y = -\frac{g_e}{\sqrt{2}} \left[(\overline{\nu}_L, \overline{e}_L) \begin{pmatrix} 0 \\ v+H \end{pmatrix} e_R + \overline{e}_R \, (0, v+H) \begin{pmatrix} \nu_L \\ e_L \end{pmatrix} \right]$$

$$= -\frac{g_e}{\sqrt{2}} v \left[\overline{e}_L e_R + \overline{e}_R e_L \right] - \frac{g_e}{\sqrt{2}} H \left[\overline{e}_L e_R + \overline{e}_R e_L \right] = -m_e \, \overline{e}e - \frac{m_e}{v} H \, \overline{e}e$$

and thus the electron mass m_e and the Yukawa coupling g_e according to

$$\boxed{m_e = \frac{g_e v}{\sqrt{2}}, \quad g_e = \frac{m_e \sqrt{2}}{v} = \frac{g_2}{\sqrt{2}} \frac{m_e}{M_W}} \tag{6.93}$$

Combined with the kinetic term in $\mathcal{L}_0$ one obtains for the e field

$$\mathcal{L}_0 + \mathcal{L}_Y |_e = \overline{e} \, i\gamma^\mu \partial_\mu e - m \, \overline{e}e - \frac{m_e}{v} H \, \overline{e}e,$$

the usual free Lagrangian for a massive Dirac field plus an interaction with the Higgs field. The extension to three lepton families is done by replication of the term in Eq. (6.92) with individual coupling constants for each case, $g_e \rightarrow g_\mu, g_\tau$, which in combination with v generate the masses $m_{\mu.\tau} = g_{\mu,\tau} v / \sqrt{2}$ and moreover define the coupling of the leptons to the Higgs field.

6.4.5.2 *Quarks*

For a single generation of quarks with masses m_u and m_d, two mass terms are required with two independent Yukawa coupling constants g_u and g_d. Expressed by u_R, d_R and the left-handed doublet

$$Q_L = \begin{pmatrix} u_L \\ d_L \end{pmatrix}, \quad \overline{Q}_L = (\overline{u}_L, \overline{d}_L),$$

the gauge invariant interaction can be written as follows,

$$\mathcal{L}_Y = -g_d \left[\overline{Q}_L \, \Phi \, d_R + \overline{d}_R \, \Phi^\dagger Q_L \right] + g_u \left[\overline{Q}_L \, \Phi^c \, u_R + \overline{u}_R \, \Phi^{c\dagger} Q_L \right]. \quad (6.94)$$

The d-part has been copied directly from the lepton sector. For m_u, however, the charge-conjugate scalar doublet is needed, $\Phi^c = -i\sigma_2 \Phi^*$, in the unitary gauge

$$\Phi^c = \frac{1}{\sqrt{2}} \begin{pmatrix} -v - H(x) \\ 0 \end{pmatrix}, \quad (6.95)$$

yielding

$$\mathcal{L}_Y = -\frac{g_d}{\sqrt{2}} (v + H) \left[\overline{d}_L d_R + \overline{d}_R d_L \right] - \frac{g_u}{\sqrt{2}} (v + H) \left[\overline{u}_L u_R + \overline{u}_R u_L \right]$$

$$= -m_d \, \overline{d} d - m_u \, \overline{u} u - \frac{m_d}{v} H \overline{d} d - \frac{m_u}{v} H \overline{u} u, \quad (6.96)$$

and accordingly the quark masses come about in analogy to the lepton masses,

$$m_u = \frac{g_u v}{\sqrt{2}}, \quad m_d = \frac{g_d v}{\sqrt{2}}. \quad (6.97)$$

In extending the concept to three quark families an essential new aspect emerges: differently from the leptons with massless neutrinos, quarks from different generations can mix. In the formalism of Yukawa interactions this feature is represented by non-diagonal matrices of coupling constants. For a compact notation multiplets of quark fields, for both L and R components, are introduced,

$$U = \begin{pmatrix} u \\ c \\ t \end{pmatrix}, \quad D = \begin{pmatrix} d \\ s \\ b \end{pmatrix}, \quad \overline{U} = (\overline{u}, \overline{c}, \overline{t}), \quad \overline{D} = (\overline{d}, \overline{s}, \overline{b}).$$

Thereby, in matrix notation, the Yukawa interaction can be written as follows,

$$\mathcal{L}_Y = - \left(\overline{U}_L, \overline{D}_L \right) G_d \, \Phi \, D_R + \left(\overline{U}_L, \overline{D}_L \right) G_u \, \Phi^c \, U_R + h.c. \quad (6.98)$$

with 3×3-matrices $G_d = (g_d^{ij})$ and $G_u = (g_u^{ij})$, with complex entries in general. Inserting $\Phi = \Phi_0$ one obtains the mass terms,

$$\mathcal{L}_{\mathrm{Y,mass}} = -\overline{D}_L \, G_d \underbrace{\frac{v}{\sqrt{2}}}_{M_d} \, D_R - \overline{U}_L \, G_u \underbrace{\frac{v}{\sqrt{2}}}_{M_u} \, U_R + h.c.$$

involving non-diagonal mass matrices M_u and M_d. Diagonalization is achieved by a transformation of all four flavour triplets

$$\begin{aligned}
U_{L,R} = V_{L,R}^u \hat{U}_{L,R}, \quad \overline{U}_{L,R} = \hat{\overline{U}}_{L,R} (V_{L,R}^u)^\dagger \\
D_{L,R} = V_{L,R}^d \hat{D}_{L,R}, \quad \overline{D}_{L,R} = \hat{\overline{D}}_{L,R} (V_{L,R}^d)^\dagger
\end{aligned} \tag{6.99}$$

into mass eigenstates $\hat{U}_{L,R}$ and $\hat{D}_{L,R}$ with the help of unitary matrices $V_{L,R}^u$ and $V_{L,R}^d$, yielding

$$\overline{D}_L M_d D_R + \overline{U}_L M_u U_R = \hat{\overline{D}}_L \underbrace{(V_L^d)^\dagger M_d V_R^d}_{M_d^{\mathrm{diag}}} \hat{D}_R + \hat{\overline{U}}_L \underbrace{(V_L^u)^\dagger M_u V_R^u}_{M_u^{\mathrm{diag}}} \hat{U}_R$$

now with diagonal mass matrices

$$M_d^{\mathrm{diag}} = \begin{pmatrix} m_d & & \\ & m_s & \\ & & m_b \end{pmatrix}, \quad M_u^{\mathrm{diag}} = \begin{pmatrix} m_u & & \\ & m_c & \\ & & m_t \end{pmatrix}.$$

After transformation into the mass eigenstates the Yukawa Lagrangian (6.98) with Φ in the unitary gauge becomes diagonal,

$$\begin{aligned}
\mathcal{L}_{\mathrm{Y}} &= -\hat{\overline{D}}_L \, M_d^{\mathrm{diag}} \, \hat{D}_R \left(1 + \frac{H}{v}\right) - \hat{\overline{U}}_L \, M_u^{\mathrm{diag}} \, \hat{U}_R \left(1 + \frac{H}{v}\right) + h.c. \\
&= -\hat{\overline{D}} \, M_d^{\mathrm{diag}} \, \hat{D} \left(1 + \frac{H}{v}\right) - \hat{\overline{U}} \, M_u^{\mathrm{diag}} \, \hat{U} \left(1 + \frac{H}{v}\right),
\end{aligned} \tag{6.100}$$

also in the residual interaction with the Higgs field H.

In order to rewrite the complete quark sector of the Standard Model Lagrangian in terms of the mass eigenstates one has to perform the transformations $U \to \hat{U}$, $D \to \hat{D}$ also in the free kinetic Dirac term and in the various interaction terms. In the kinetic part as well as in the electromagnetic and NC interactions exclusively flavour-diagonal terms are present, because of

$$\overline{U}_{L(R)} \gamma^\mu (\gamma_5) U_{L(R)} = \hat{\overline{U}}_{L(R)} \gamma^\mu (\gamma_5) \hat{U}_{L(R)},$$

$$\overline{D}_{L(R)} \gamma^\mu (\gamma_5) D_{L(R)} = \hat{\overline{D}}_{L(R)} \gamma^\mu (\gamma_5) \hat{D}_{L(R)},$$

owing to the unitarity of the matrices $V_{L,R}^{u,d}$, thus retaining the original form also when mass eigenstates are used. The only item that is modified by the transformation is the CC interaction with the charged current

$$\overline{U}_L\gamma^\mu D_L + \overline{D}_L\gamma^\mu U_L = \overline{\hat{U}}_L V \gamma^\mu \hat{D}_L + \overline{\hat{D}}_L V^\dagger \gamma^\mu \hat{U}_L,$$

which now includes as a remant the unitary matrix

$$V \equiv V_{\text{CKM}} = (V_L^u)^\dagger V_L^d. \tag{6.101}$$

This matrix is known as the CKM matrix, named after *Cabbibo, Kobayashi, and Maskawa*. Originally flavor mixing was introduced by Cabbibo for the charged-current interaction of hadrons with strangeness [Cabibbo (1963)]; with the rise of the quark model, it could be assigned to mixing between the first two quark generations. Later the concept of quark mixing was extended to three generations by Kobayashi and Maskawa in 1973 [Kobayashi (1973)].

In the following we agree on making use of the mass eigenstates everywhere for the fields and their interactions, without the special notation, i.e. leaving aside the superscript ˆ on the quark fields. Thus, $\mathcal{L}_{\text{int}}$ remains unchanged for the electromagnetic and NC interactions; the CC interactions turn into

$$\mathcal{L}_{\text{int}}^{(CC)} = \frac{g_2}{\sqrt{2}} \left[\sum_{\ell=e,\mu,\tau} \overline{\nu}_\ell \gamma^\mu \frac{1-\gamma_5}{2} \ell\, W_\mu^+ + \sum_{i,j} \overline{u}^i \gamma^\mu \frac{1-\gamma_5}{2} V_{ij}\, d^j\, W_\mu^+ + h.c. \right] \tag{6.102}$$

making use of the notations $u^i = u,c,t$ and $d^i = d,s,b$ together with

$$V_{\text{CKM}} = (V_{ij}) = \begin{pmatrix} V_{ud} & V_{us} & V_{ub} \\ V_{cd} & V_{cs} & V_{cb} \\ V_{td} & V_{ts} & V_{tb} \end{pmatrix}. \tag{6.103}$$

As a complex 3×3 matrix, V_{CKM} has 18 real entries, which are reduced to 9 by the unitarity conditions. Out of them, 5 phases can be absorbed in the quark fields, leaving a single complex phase. Thus, V_{CKM} is parametrized by three angles $\theta_{12}, \theta_{13}, \theta_{23}$ and one phase δ. These four quantitites are additional free parameters of the Standard Model and have to be determined by experiment. The presence of the complex phase gives rise to CP violation in the weak interaction. It is interesting to note that with less than three generations no complex phase and thus no CP violation occurs, which makes three fermion generations in a certain sense unique: as the minimal texture for the presence of CP violation.

Note on neutrino masses. In the Standard Model, neutrino fields are left-handed and neutrinos are treated as massless. It is, nevertheless, a straight-forward extension to add a right-handed neutrino field ν_R, which is a gauge singlet, and to introduce a mass term in analogy to that in Eq. (6.94) for u quarks to obtain a non-zero neutrino mass. Furthermore, an additional Majorana mass term $M\overline{\nu_R^c}\,\nu_R$ is possible involving the charge-conjugate of the ν_R field. For three lepton families, neutrinos can mix in analogy to mixing in the quark sector, allowing basically also CP-violation in the lepton sector. Whereas neutrino mixing is meanwhile experimentally established, there is no indication as yet for leptonic CP violation.

6.5 Phenomenology and Tests of the Standard Model

The predictive power of the electroweak Standard Model offers a multitude of possibilities for tests, ranging from the low-energy region of particle decays and neutrino scattering to the currently highest energies at the Large Hadron Collider LHC.

6.5.1 *Neutral currents and neutrino scattering*

The neutral current interaction facilitates further neutrino scattering reactions that are mediated by the exchange of neutral Z bosons and hence are named neutral-current processes. They are a prediction of the Standard Model, and therefore their discovery in 1973 [Hasert (1973)] has been a first great confirmation, ten years before the discovery of the vector bosons $W^\pm$ and Z.

6.5.1.1 *Neutrino–electron scattering*

In Sec. 6.3.3 the elastic scattering of muon neutrinos by electrons via the charged current was treated. Now we consider ν_μ scattering by electrons via the neutral current described by the Feynman diagram with Z^0 exchange,

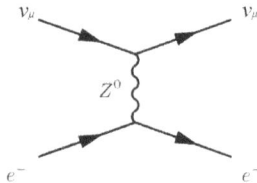

corresponding to the matrix element $\mathcal{M}$ which contains the NC couplings (6.62) and the Z propagator in the low-energy approximation, following Sec. 6.3.3. Adopting the kinematical notations of Sec. 6.3.3 one obtains the spin-averaged squared matrix element,

$$
\overline{|\mathcal{M}|^2} = \left(\frac{g_2^2}{4c_W^2 M_Z^2}\right)^2 \cdot 8s^2 \left[(v_e + a_e)^2 + (v_e - a_e)^2(1 - y)^2\right]
$$
$$
= 16\, G_F^2\, s^2 \left[(v_e + a_e)^2 + (v_e - a_e)^2(1 - y)^2\right], \qquad (6.104)
$$

where the last step made use of the relation (6.86) between the mixing angle and the vector boson masses. The differential cross section

$$
\frac{d\sigma}{dQ^2}(\nu_\mu e^-) = \frac{1}{16\pi s^2}\, \overline{|\mathcal{M}|^2}
$$

yields after integration

$$
\sigma(\nu_\mu e^-) = \frac{G_F^2}{4\pi} s \left[(v_e + a_e)^2 + \frac{1}{3}(v_e - a_e)^2\right]. \qquad (6.105)
$$

For the antineutrino scattering cross section one obtains accordingly

$$
\sigma(\bar{\nu}_\mu e^-) = \frac{G_F^2}{4\pi} s \left[(v_e - a_e)^2 + \frac{1}{3}(v_e + a_e)^2\right]. \qquad (6.106)
$$

The ratio of the two cross sections,

$$
R_{\nu e} = \frac{1 + \xi + \xi^2}{1 - \xi + \xi^2}, \quad \xi = \frac{v_e}{a_e} = 1 - 4\sin^2\theta_W, \qquad (6.107)
$$

provides a possibility for the experimental determination of $\sin^2\theta_W$. The result of the most precise measurements [Vilain (1994)] is given by

$$
\sin^2\theta_W = 0.2324 \pm 0.0083. \qquad (6.108)
$$

6.5.1.2 *Neutrino–nucleon scattering*

The neutral current processes in deep-inelastic neutrino scattering yield additional contributions to the nucleon structure functions and, moreover, provide further possibilities for measuring the electroweak mixing angle. The elementary parton processes consist in scattering of neutrinos by the quarks and antiquarks of the nucleon. Since there is no difference between ν_e and ν_μ cross sections, the simpler notation ν will be used.

In case of $\nu q \to \nu q$ scattering, the matrix element is obtained from that of $\nu_\mu e^- \to \nu_\mu e^-$ replacing v_e, a_e by the couplings v_q, a_q, and thus the cross section follows immediately from Eq. (6.105),

$$\sigma^{\nu q} = \frac{G_F^2}{4\pi} s \left[(v_q + a_q)^2 + \frac{1}{3}(v_q - a_q)^2 \right]. \tag{6.109}$$

The cross section for neutrino scattering by antiquarks, $\nu \bar{q} \to \nu \bar{q}$, has the same form as Eq. (6.106), with the respective quark coupling constants,

$$\sigma^{\nu \bar{q}} = \frac{G_F^2}{4\pi} s \left[(v_q - a_q)^2 + \frac{1}{3}(v_q + a_q)^2 \right]. \tag{6.110}$$

The differential cross section for neutrino scattering by protons and neutrons follows from the partonic cross sections by multiplication with the quark distribution functions and summation over all contributing quarks and antiquarks. Neglecting the sea quarks of the second and third generation one obtains

$$\frac{d\sigma}{dx}(\nu P) = \sigma^{\nu u} u(x) + \sigma^{\nu \bar{u}} \bar{u}(x) + \sigma^{\nu d} d(x) + \sigma^{\nu \bar{d}} \bar{d}(x),$$

$$\frac{d\sigma}{dx}(\nu N) = \sigma^{\nu u} d(x) + \sigma^{\nu \bar{u}} \bar{d}(x) + \sigma^{\nu d} u(x) + \sigma^{\nu \bar{d}} \bar{u}(x). \tag{6.111}$$

For an isoscalar target the cross section follows by averaging,

$$\frac{d\sigma}{dx}(\nu \frac{P+N}{2}) = \frac{1}{2}(\sigma^{\nu u} + \sigma^{\nu d})\left[u(x) + d(x)\right] + \frac{1}{2}(\sigma^{\nu \bar{u}} + \sigma^{\nu \bar{d}})\left[\bar{u}(x) + \bar{d}(x)\right]. \tag{6.112}$$

The integration over x, taking into account $s = xS$ for the partonic variable s, can be shifted to the parton distribution functions, yielding the integrated distributions

$$U + D = \int_0^1 dx\, x\left[u(x) + d(x)\right], \quad \overline{U} + \overline{D} = \int_0^1 dx\, x\left[\bar{u}(x) + \bar{d}(x)\right]. \tag{6.113}$$

The NC cross section for an isoscalar target can thus be written in the following way,

$$\sigma_{NC}^\nu = \frac{G_F^2}{2\pi} S \left(\left(\frac{1}{2} - s_W^2 + \frac{5}{9}s_W^4 \right) \left[U + D + \frac{1}{3}(\overline{U} + \overline{D}) \right] \right.$$

$$\left. + \frac{5}{9}s_W^4 \left[\overline{U} + \overline{D} + \frac{1}{3}(U + D) \right] \right). \tag{6.114}$$

Accordingly, the integrated CC cross sections for an isoscalar target follow from Eqs. (6.32) and (6.33) for neutrino and antineutrino scattering,

$$\sigma^{\bar{\nu}}_{CC} = \frac{G_F^2}{2\pi} S \left[U + D + \frac{1}{3}(\overline{U} + \overline{D}) \right],$$

$$\sigma^{\nu}_{CC} = \frac{G_F^2}{2\pi} S \left[\overline{U} + \overline{D} + \frac{1}{3}(U + D) \right]. \tag{6.115}$$

From the ratio of the NC/CC cross sections, information on electroweak parameters can be gained widely independent of the properties of the nucleon. An important role is played by the quantity

$$R_\nu = \frac{\sigma^{\nu}_{NC}}{\sigma^{\nu}_{CC}} = \frac{1}{2} - s_W^2 + (1 + r)\frac{5}{9}s_W^4, \quad \text{with } r = \frac{\sigma^{\bar{\nu}}_{CC}}{\sigma^{\nu}_{CC}} \tag{6.116}$$

for the determination of s_W^2. Numerically one has $r \simeq 0.5$, corresponding to the ratio $(\overline{U}+\overline{D})/(U+D) = 0.2$. For antineutrinos there is an analogous quantity,

$$R_{\bar{\nu}} = \frac{\sigma^{\bar{\nu}}_{NC}}{\sigma^{\bar{\nu}}_{CC}} = \frac{1}{2} - s_W^2 + \left(1 + \frac{1}{r}\right)\frac{5}{9}s_W^4, \tag{6.117}$$

moreover, the combination

$$R^- = \frac{\sigma^{\nu}_{NC} - \sigma^{\bar{\nu}}_{NC}}{\sigma^{\nu}_{CC} - \sigma^{\bar{\nu}}_{CC}} \tag{6.118}$$

is special because of low systematic uncertainties. From precise measurements of R_ν and $R_{\bar{\nu}}$ from 1984 until 1998 the values in Table 6.3 were extracted, combined by averaging as follows,

$$s_W^2 = 0.2276 \pm 0.0030. \tag{6.119}$$

This value has to be compared to the one from the mass ratio,

$$s_W^2 = 1 - \frac{M_W^2}{M_Z^2} = 0.2230 \pm 0.0002 \tag{6.120}$$

Table 6.3 s_W^2 measured in νN scattering.

Experiment	s_W^2
CHARM [Allaby (1987)]	0.236 ± 0.006
CDHS [Blondel (1990)]	0.228 ± 0.006
CCFR [McFarland (1998)]	0.2236 ± 0.0041

with the measured masses of W and Z (see next section); both results are compatible within two standard deviations. The most precise measurement via the ratio R^- is from the year 2002 by the NuTeV Collaboration [Zeller (2002)]. The extracted value

$$s_W^2 = 0.2277 \pm 0.0016,$$

however, is three standard deviations above the result in Eq. (6.120). The deviation is still under debate, a conclusion has not been reached.

6.5.2 *The vector boson masses*

The vector bosons $W^\pm$ und Z were discovered 1983 [Arnison (1983); Banner (1983)] at the SPS, the Super Proton Synchrotron at CERN, produced by quark–antiquark annihilation in proton–antiproton collisions and detected via their subsequent decays into lepton pairs,

$$Z \to \ell^+\ell^-, \quad W^+ \to \ell^+\nu_\ell, \quad W^- \to \ell^-\bar\nu_\ell \quad (\ell = e, \mu)$$

from which also the masses M_W and M_Z were reconstructed for the first time. The most precise determination of the Z-boson mass took place later at the electron–positron collider LEP at CERN during the first phase (LEP I) from 1989 until 1995. Precise measurements of the W-boson mass followed afterwards during the second phase of LEP (LEP II) from 1995 until 2000. Simultaneously they were carried out at the proton–antiproton collider TEVATRON where they continued until 2011 when the TEVATRON was decommissioned. In 2010 the experiments at the LHC took over. The experimental results for the vector boson masses are summarized by the averaged values [Zyla (2020)]

$$
\begin{aligned}
M_W &= 80.379 \pm 0.012 \,\text{GeV}, \\
M_Z &= 91.1876 \pm 0.0021 \,\text{GeV}.
\end{aligned}
\tag{6.121}
$$

The proton–proton collisions at the LHC will continue to produce W and Z bosons and thus will contribute to a further improvement of the W mass determination, aiming at an uncertainty of 10 MeV or even less.

Figure 6.2 shows the events for Z-boson production with subsequent decay into muon pairs, $Z \to \mu^+\mu^-$, as measured at the LHC by the CMS experiment [Chatrchyan (2012a)], together with neutral mesons composed of u, d, s, c, b quarks, displayed over the full energy range covered at a total energy $\sqrt{S} = 7\,\text{TeV}$.

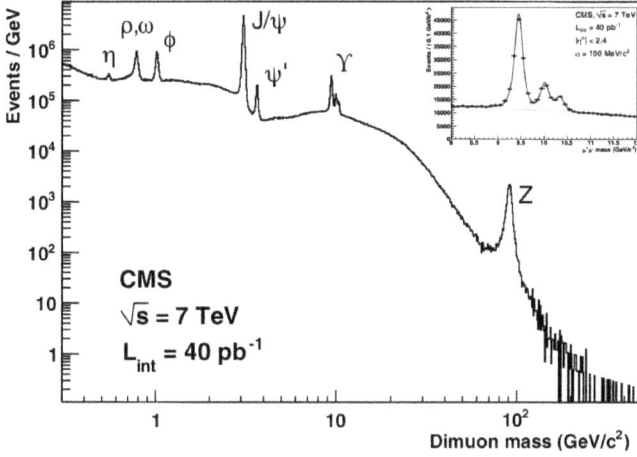

Fig. 6.2 Event numbers for $PP \to \mu^+\mu^- X$ at the LHC versus the invariant mass $m_{\mu^+\mu^-}$.

The following example illustrates how the cross sections for W and Z production in hadron collisions are computed.

Applying the Feynman rules of Sec. 6.4.2, the matrix elements for $q\bar{q}' \to W^\pm$ and $q\bar{q} \to Z$ read as follows (including the CKM matrix element $V_{qq'}$)

$$\mathcal{M}_W = \frac{g_2}{2\sqrt{2}} V_{qq'} \, \bar{v}(p_2)\gamma^\mu(1-\gamma_5)u(p_1) \, \epsilon_\mu(k),$$

$$\mathcal{M}_Z = \frac{g_2}{2c_W} \bar{v}(p_2)\gamma^\mu(v_q - a_q\gamma_5)u(p_1) \, \epsilon_\mu(k).$$

With the standard methods and the polarization sum (6.34) for massive vector bosons one obtains

$$\overline{|\mathcal{M}_W|^2} = \left\langle \frac{1}{3} \right\rangle \frac{g_2^2}{4} |V_{qq'}|^2 M_W^2,$$

$$\overline{|\mathcal{M}_Z|^2} = \left\langle \frac{1}{3} \right\rangle \frac{g_2^2}{4c_W^2} \left(v_q^2 + a_q^2\right) M_Z^2;$$

the factor $1/3$ originates from averaging over the colour states of the incoming quarks and antiquarks. The partonic cross section for $V = W, Z$ follow from the general expression given in Eq. (3.72) with $s = (p_1 + p_2)^2$,

$$d\hat{\sigma}_V = \frac{\pi}{s} \overline{|\mathcal{M}_V|^2} \, \delta^4(p_1 + p_2 - k) \frac{d^3k}{2k_0}$$

$$= \frac{\pi}{s} \overline{|\mathcal{M}_V|^2} \, \delta^4(p_1 + p_2 - k) \, \delta(k^2 - M_V^2) \, d^4k$$

and integration over $d^4 k$ yields

$$\hat{\sigma}_V(s) = \frac{\pi}{s}\, \overline{|\mathcal{M}_V|^2}\, \delta(s - M_V^2).$$

In detail, one obtains for $\hat{\sigma}_{q\bar{q}' \to W}(s) \equiv \hat{\sigma}_W(s)$ and $\hat{\sigma}_{q\bar{q}' \to Z}(s) \equiv \hat{\sigma}_Z(s)$:

$$\hat{\sigma}_W(s) = \pi\, \frac{g_2^2}{12}\, |V_{qq'}|^2\, \delta(s - M_W^2) = \pi\, \frac{\sqrt{2}}{3}\, G_F M_W^2\, |V_{qq'}|^2\, \delta(s - M_W^2),$$

$$\hat{\sigma}_Z(s) = \pi\, \frac{g_2^2}{12\, c_W^2}\, (v_q^2 + a_q^2)\, \delta(s - M_Z^2) = \pi\, \frac{\sqrt{2}}{3}\, G_F M_Z^2\, (v_q^2 + a_q^2)\, \delta(s - M_Z^2).$$

The hadronic production cross sections for $h_1 h_2 \to W, Z$ in proton–antiproton ($h_1 = P, h_2 = \overline{P}$) and in proton–proton ($h_1 = h_2 = P$) collisions follow from the partonic cross sections by multiplication with the parton distribution functions and subsequent integration over the parton momenta, as described in Sec. 5.2. From $p_1 = x_1 P_1$, $p_2 = x_2 P_2$ with the notation P_1 and P_2 for the momenta of h_1 und h_2 one obtains $s = x_1 x_2 S$ with $S = (P_1 + P_2)^2$, and the hadronic cross sections can be written as follows,

$$\sigma_Z = \sum_q \int_0^1 dx_1 \int_0^1 dx_2 \left[f_q^{h_1}(x_1)\, f_{\bar{q}}^{h_2}(x_2) + f_{\bar{q}}^{h_1}(x_1)\, f_q^{h_2}(x_2) \right] \hat{\sigma}_{q\bar{q} \to Z}(x_1 x_2 S)$$

$$= \sum_q \int_0^1 d\tau\, \frac{d\mathcal{L}_{q\bar{q}}}{d\tau}(\tau)\, \hat{\sigma}_{q\bar{q} \to Z}(\tau S) = \frac{\pi}{3} \sqrt{2}\, G_F \sum_q (v_q^2 + a_q^2)\, \frac{d\mathcal{L}_{q\bar{q}}}{d\tau}(\tau_Z),$$

$$\sigma_W = \sum_{q,\bar{q}'} \int_0^1 dx_1 \int_0^1 dx_2 \left[f_q^{h_1}(x_1) f_{\bar{q}'}^{h_2}(x_2) + f_{\bar{q}'}^{h_1}(x_1) f_q^{h_2}(x_2) \right] \hat{\sigma}_{q\bar{q}' \to W}(x_1 x_2 S)$$

$$= \sum_{q,\bar{q}'} \int_0^1 d\tau\, \frac{d\mathcal{L}_{q\bar{q}'}}{d\tau}(\tau)\, \hat{\sigma}_{q\bar{q}' \to W}(\tau S) = \frac{\pi}{3} \sqrt{2}\, G_F \sum_{q,\bar{q}'} |V_{qq'}|^2\, \frac{d\mathcal{L}_{q\bar{q}'}}{d\tau}(\tau_W)$$

using the notations $\tau_Z = M_Z^2/S$, $\tau_W = M_W^2/S$. Therein, the following parton luminosities were introduced,

$$\frac{d\mathcal{L}_{q\bar{q}}}{d\tau}(\tau) = \int_\tau^1 \frac{dx}{x} \left[f_q^{h_1}(x)\, f_{\bar{q}}^{h_2}\left(\frac{\tau}{x}\right) + f_{\bar{q}}^{h_1}(x)\, f_q^{h_2}\left(\frac{\tau}{x}\right) \right],$$

$$\frac{d\mathcal{L}_{q\bar{q}'}}{d\tau}(\tau) = \int_\tau^1 \frac{dx}{x} \left[f_q^{h_1}(x)\, f_{\bar{q}'}^{h_2}\left(\frac{\tau}{x}\right) + f_{\bar{q}'}^{h_1}(x)\, f_q^{h_2}\left(\frac{\tau}{x}\right) \right].$$

The scale Q^2 in the parton distributions has been suppressed in the notations above. It has to be chosen as $Q^2 = M_W^2$ or $Q^2 = M_Z^2$, respectively.

6.5.2.1 *Muon decay and the vector-boson mass correlation*

An important correlation between the masses M_W and M_Z is achieved via the decay width of the muon calculated in the Standard Model and identified with the result of the Fermi model in the low-energy limit. Such a step was already done in the $V-A$ theory with the result (6.27); it agrees with that of the Standard Model when g is identified with the $SU(2)$ coupling constant g_2, hence

$$\frac{g_2^2}{8M_W^2} = \frac{G_F}{\sqrt{2}}. \tag{6.122}$$

Inserting the representation (6.85) for M_W allows to determine the vacuum expectation value of the Higgs field,

$$\frac{1}{v^2} = \sqrt{2}\,G_F \Rightarrow v = \left(\sqrt{2}\,G_F\right)^{-1/2} = 246\,\text{GeV}. \tag{6.123}$$

Moreover, the connection (6.60) between e and g_2 yields the relation

$$\frac{g_2^2}{8M_W^2} = \frac{e^2}{8M_W^2 s_W^2} = \frac{\pi\alpha}{2M_W^2 s_W^2} = \frac{G_F}{\sqrt{2}}.$$

Historically this made a first prediction of the vector boson masses possible by means of s_W^2 extracted from neutrino scattering. With the relation (6.86) one obtains the $W-Z$ mass correlation,

$$\boxed{M_W^2 \left(1 - \frac{M_W^2}{M_Z^2}\right) = \frac{\pi\alpha}{\sqrt{2}\,G_F} = \left(37.28038(1)\,\text{GeV}\right)^2}. \tag{6.124}$$

Taking the precisely measured Z mass together with α and G_F as input quantities, the W mass is predicted as follows,

$$M_W = 80.939 \pm 0.002\,\text{GeV},$$

to be compared to the measured value of M_W in Eq. (6.121). The prediction for the mass ratio,

$$\frac{M_W}{M_Z} = 0.8876$$

has to be compared to the value from the directly measured masses in Eq. (6.121),

$$\left(\frac{M_W}{M_Z}\right)_{\text{exp}} = 0.88147 \pm 0.00015. \tag{6.125}$$

One recognizes a significant deviation between the theory predictions and the experimental results. This "discrepancy", however, is by no means a deficit of the Standard Model, but just the opposite: the measurements are so precise that the perturbative computation of the muon decay width at lowest order is not sufficient and higher orders with loop contributions to the decay amplitude play an essential role. The detected deviation is actually a confirmation of the existence of the loop contributions; the quantitative agreement of the calculated corrections with the measured values constitutes a successful precision test of the Standard Model, in analogy to the anomalous magnetic moment of the electron as a precision test of QED (see Sec. 3.8). More details will be given in Sec. 6.5.4.

6.5.3 *Vector bosons in electron–positron annihilation*

The most precise measurements of the Z-boson properties were done by electron–positron annihilation at the e^+e^- collider LEP at CERN and at the Stanford Linear Collider (SLC) at SLAC. Among the standard reactions are fermion-pair production processes $e^+e^- \to \gamma, Z \to f\bar{f}$ via photon and Z^0 exchange,

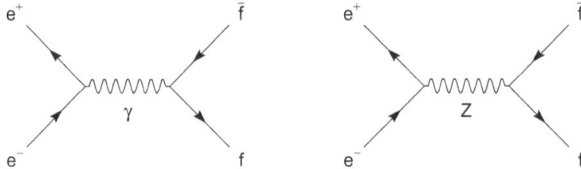

In Sec. 3.7 the cross section for $e^+e^- \to \gamma \to f\bar{f}$ was calculated within QED. Above 30 GeV the description of the cross section by QED becomes more and more insufficient, in the range around 90 GeV the cross section is almost completely dominated by the weak interaction. This is illustrated in Fig. 6.3 showing the example of hadron production, which proceeds via primary quark–antiquark pairs according to $e^+e^- \to q\bar{q} \to$ hadrons.

6.5.3.1 *Z resonance*

As an unstable particle the Z boson appears in the cross section as a resonance with a finite width Γ_Z, manifesting itself in the propagator in terms of an imaginary part in the denominator. The expression in Eq. (6.26) (with $s = q^2$) has to be modified accordingly,

$$\frac{1}{s - M_Z^2 + i\epsilon} \to \frac{1}{s - M_Z^2 + iM_Z\Gamma_Z} \equiv D_Z(s) \qquad (6.126)$$

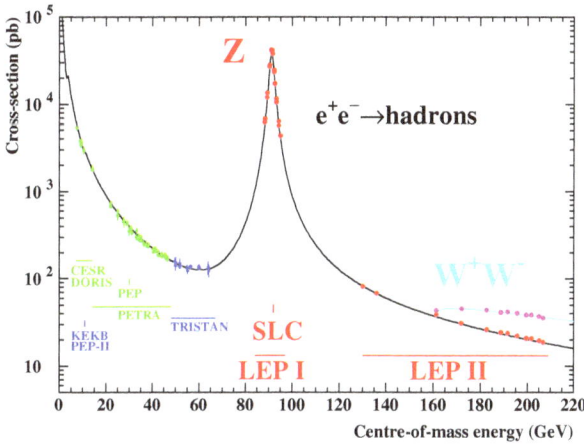

Fig. 6.3 Cross section for $e^+e^- \to$ hadrons, theory prediction and measurements [Schael (2005)].

which generates a Breit-Wigner shape of the cross section,

$$\sigma_Z(s) \sim \frac{1}{(s - M_Z^2)^2 + M_Z^2 \Gamma_Z^2}.$$

The width is calculated according to

$$\Gamma_Z = \sum_f \Gamma_f, \quad \Gamma_f = \Gamma(Z \to f\bar{f}) \tag{6.127}$$

as the sum over the partial widths for the decays $Z \to f\bar{f}$ into all fermions with $2m_f < M_Z$. The calculation of the partial widths is based on the matrix element $\mathcal{M}(Z \to f\bar{f})$ with the couplings in Eq. (6.64) and the general formula (3.83), yielding (fermion masses neglected)

$$\Gamma_f = \frac{1}{16\pi M_Z} \overline{|\mathcal{M}|^2} = \frac{1}{16\pi M_Z} \cdot N_C^f \frac{g_2^2}{3c_W^2} M_Z^2 \left(v_f^2 + a_f^2\right)$$

$$= N_C^f \frac{G_F M_Z^3}{6\pi\sqrt{2}} \left(v_f^2 + a_f^2\right) \tag{6.128}$$

with the colour factor

$$N_C^f = \begin{cases} 3 & \text{for } f = \text{quarks} \\ 1 & \text{for } f = \text{leptons}. \end{cases}$$

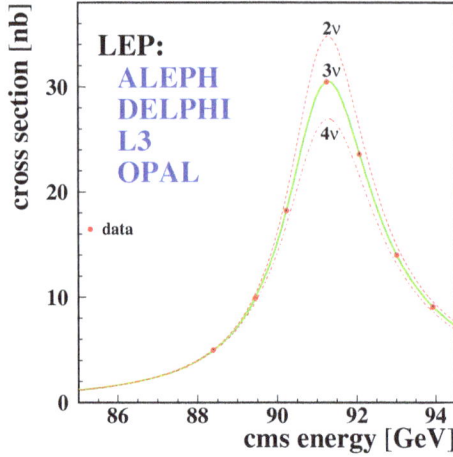

Fig. 6.4 Measurement of the lineshape around the Z resonance [Schael (2005)]. The curves indicate the predicted cross section for two, three and four neutrino species with zero mass.

From measurements of the line shape, i.e. the s-dependence of the cross section, the mass M_Z has been obtained as given in Eq. (6.121), together with the width Γ_Z as follows [Schael (2005)],

$$\Gamma_Z^{\text{exp}} = 2.4952 \pm 0.0023 \, \text{GeV}. \tag{6.129}$$

The measured line shape is compatible with the predicted curve only for $N_\nu = 3$ for the number of neutrinos (see Fig. 6.4), Historically this was the first direct confirmation for the existence of exactly three fermion families (with light neutrinos). Further comparisons of theoretical predictions and experimental results will be discussed in the next section.

For calculating the resonance cross section for $e^+e^- \to f\bar{f}$ it is sufficient to consider only the Z-exchange diagram in the matrix element, given by

$$\mathcal{M}_Z = i \, \frac{g_2^2}{4c_W^2} \, D_Z(s) \left[\bar{v}(p_2)\gamma^\mu(v_e - a_e\gamma_5)u(p_1) \right] \left[\bar{u}(p_3)\gamma_\mu(v_f - a_f\gamma_5)v(p_4) \right]$$

with $s = (p_1 + p_2)^2$ and the complex propagator D_Z in Eq. (6.126). The unpolarized cross section in the CMS (neglecting fermion masses),

$$\frac{d\sigma_Z^f}{d\Omega} = \frac{1}{64\pi^2 s} \, \overline{|\mathcal{M}_Z|^2},$$

is derived applying the rules of Sec. 3.7 and Sec. 6.3.2, with the result

$$
\frac{d\sigma_Z^f}{d\Omega} = \frac{N_C}{64\pi^2} \frac{s}{(s - M_Z^2)^2 + M_Z^2 \Gamma_Z^2} \left(\frac{g_2^2}{4c_W^2}\right)^2 \left[A\left(1 + \cos^2\theta\right) + B\cdot 2\cos\theta\right]
$$

$$
= \frac{N_C}{64\pi^2} \frac{s}{(s - M_Z^2)^2 + M_Z^2 \Gamma_Z^2} \left(\sqrt{2}G_F M_Z^2\right)^2 \left[A\left(1 + \cos^2\theta\right) + B\cdot 2\cos\theta\right]
$$

$$(6.130)$$

containing the specific combinations of coupling constants

$$
A = (v_e^2 + a_e^2)(v_f^2 + a_f^2) \quad \text{and} \quad B = (2v_e a_e)(2v_f a_f).
$$

Integration yields the resonance cross section in the Breit-Wigner representation,

$$
\boxed{\sigma_Z^f(s) = \sigma_0^f \frac{s\,\Gamma_Z^2}{(s - M_Z^2)^2 + M_Z^2 \Gamma_Z^2}, \qquad \sigma_0^f = \frac{12\pi}{M_Z^2} \cdot \frac{\Gamma_e \Gamma_f}{\Gamma_Z^2}}
$$

$$(6.131)$$

where instead of the couplings the partial widths for $Z \to e^+ e^-$ and $Z \to f\bar{f}$ according to Eq. (6.128) have been introduced. This representation is distinguished by a broader validity; from a fit to the measured line shape the parameters $M_Z, \Gamma_Z, \sigma_0^f$ can be extracted in a fairly model independent way. Of particular interest are the hadronic peak cross section,

$$
\sigma_0^{\text{had}} = \sum_{q=u,\dots b} \sigma_0^q
$$

$$(6.132)$$

and the ratios

$$
\frac{\sigma_0^{\text{had}}}{\sigma_0^e} = \frac{\Gamma_{\text{had}}}{\Gamma_e} \equiv R_{\text{had}}, \qquad \frac{\Gamma_b}{\Gamma_{\text{had}}} \equiv R_b
$$

$$(6.133)$$

with the hadronic width $\Gamma_{\text{had}} = \sum_q \Gamma_q$, for tests of the Standard Model predictions.

6.5.3.2 *On-resonance asymmetries*

A specific feature of the differential cross section (6.130) is the appearance of an asymmetry in the angular distribution which becomes manifest after integration over the forward and backward regions

$$
\sigma_F = \int_{\theta < \pi/2} d\Omega\, \frac{d\sigma_Z^f}{d\Omega}, \qquad \sigma_B = \int_{\theta > \pi/2} d\Omega\, \frac{d\sigma_Z^f}{d\Omega}
$$

in a *forward-backward asymmetry*

$$A_{\text{FB}}^f = \frac{\sigma_F - \sigma_B}{\sigma_F + \sigma_B} = \frac{3}{4} A_e A_f. \tag{6.134}$$

The combinations of vector and axial vector couplings

$$A_f = \frac{2\,v_f a_f}{v_f^2 + a_f^2} = \frac{2\,(1 - 4|Q_f|\,s_W^2)}{1 + (1 - 4|Q_f|\,s_W^2)^2} \tag{6.135}$$

depend only on s_W^2, hence measurements of A_{FB} allow a precise determination of the electroweak mixing angle.

Another kind of asymmetry originates from parity violation of the weak interaction: the *left-right asymmetry*

$$A_{\text{LR}} = \frac{\sigma_L - \sigma_R}{\sigma_L + \sigma_R} \tag{6.136}$$

of the integrated cross sections for left-handed and right-handed polarized electrons, where σ_L refers to helicity $-1/2$ and σ_R to helicity $+1/2$. Since helicity and chirality are practically equal for high-energetic fermions (see Sec. 6.2), the polarized cross sections can be computed replacing in the matrix element $\mathcal{M}_Z$ the e^- spinor $u(p_1)$ by

$$u(p_1) \rightarrow \frac{1}{2}(1 \pm \gamma_5)\,u(p_1) \tag{6.137}$$

and skipping the factor $1/2$ when summing over the e^- helicities. The calculation is an easy exercise, yielding

$$A_{\text{LR}} = A_e = \frac{2\,(1 - 4s_W^2)}{1 + (1 - 4s_W^2)^2} \tag{6.138}$$

indendent of the fermions in the final state. This feature and the almost linear dependence of the quantity A_e on s_W^2 imply the most precise measurement of s_W^2 via a single observable. Such a measurement was facilitated at SLAC by the availability of a longitudinally polarized electron beam at the SLC [Abe (1994)].

The *polarisation* of the τ lepton in $e^+e^- \rightarrow \tau^+\tau^-$ belongs to the same class of asymmetries as A_{LR}, both the differential polarization

$$P_\tau(\cos\theta) = \frac{d\sigma_+ - d\sigma_-}{d\sigma_+ + d\sigma_-} \tag{6.139}$$

as well as the averaged polarization

$$\langle P_\tau \rangle = \frac{\sigma_+ - \sigma_-}{\sigma_+ + \sigma_-}, \tag{6.140}$$

defined by the cross sections with τ^- helicity $+1/2$ (σ_+) and helicity $-1/2$ (σ_-). Technically they can be computed from $\mathcal{M}_Z$ with the replacement of the τ^- spinor $u(p_3)$ by the corresponding left- and right-handed spinors, in analogy to Eq. (6.137) for the e^- spinor in the calculation of A_{LR}. One finds

$$P_\tau(\cos\theta) = -\frac{A_\tau\,(1+\cos^2\theta) + 2A_e\,\cos\theta}{1+\cos^2\theta + 2A_eA_\tau\,\cos\theta}. \tag{6.141}$$

The quantitites A_τ and A_e can be determined by a fit to the measured angular distribution. From the average τ polarization one obtains directly

$$\langle P_\tau \rangle = -A_\tau. \tag{6.142}$$

In case of lepton universality the relation $A_\tau = A_e$ holds. Independent measurements thus provide a test of the equality of the V and A couplings for the different generations, as formulated in the Standard Model.

Experimental results for the Z-boson observables and comparisons with the Standard Model predictions will be presented in the next section.

6.5.3.3 *W-boson pair production*

At CMS energies $> 160\,\mathrm{GeV}$ W-boson pairs are produced, $e^+e^- \to W^+W^-$, described by the diagrams with neutrino exchange in the t-channel together with photon and Z exchange in the s-channel,

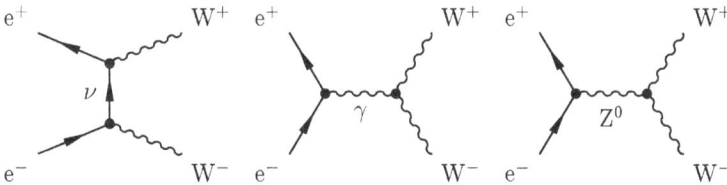

which contain the trilinear self couplings of the vector fields. In the V–A theory only the neutrino-exchange contribution exists, provoking an uncontrollable increase of the cross section with energy, as shown in Sec. 6.3.5. The completion of the amplitude by photon and Z exchange is decisive for the compensation of the increase; it has been confirmed by measurements of the energy dependence of the cross section, displayed in Fig. 6.5. Each deviation of the trilinear couplings from the gauge coupling structure given in Eq. (6.67) generates a steeper increase of the cross section. Hence, by the precise experiments severe constraints on possible "anomalous couplings" have become available.

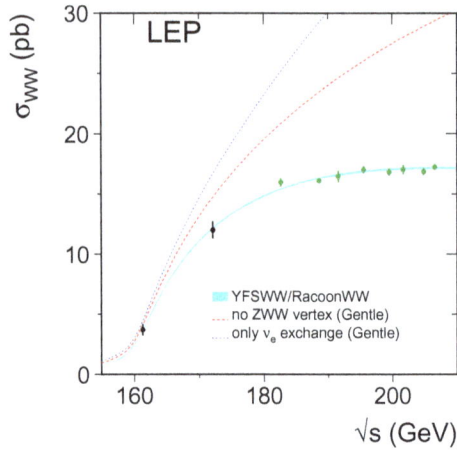

Fig. 6.5 Measured W-pair production cross section compared to the Standard Model prediction (full curve) [Schael (2013)].

Further important measurements on W-pair production have to be mentioned.

- Determination of the mass M_W of the W bosons from the energy dependence of the production cross section around the kinematical threshold.
- Reconstruction of the W mass from the decays $W \rightarrow f\bar{f}'$ into pairs of fermions, leptons as well as quarks.

Both kinds of measurements done at LEP (phase II) have contributed substantially to the precise value of M_W in Eq. (6.121).

6.5.4 *Electroweak precision tests*

The last three decades represent a period of extraordinary progress in the experimental knowledge of electroweak phenomena, with data collections of highest accuracy from the e^+e^- and hadron collider experiments of the past as well as from the ongoing LHC experiments.

The lowest-order relations between the various observables derived from the Standard Model turn out to be in general insufficient when confronted with the experimental data and require the inclusion of terms beyond the lowest order in perturbation theory. This was already made explicit for the vector boson masses $M_{W,Z}$ and their interdependence as a concrete example. By the high experimental precision the observables become sensitive to the quantum structure of the theory which appears in terms of higher-order contributions involving diagrams with closed loops in the Feynman-graph expansion, similar to the situation in QED addressed in Sec. 3.8. To visualize the quantitative relations: the typical size of loop contributions is of the order $\mathcal{O}(10^{-2})$ relative to the lowest-order terms, whereas the relative accuracy of the measurements is of $\mathcal{O}(10^{-3})$ or even better. The set of quantities presented in the previous two subsections is usually labeled by the keyword *electroweak precision observables*.

The possibility to perform perturbative calculations for accurate predictions of measureable quantitities is due to the electroweak Standard Model as a renormalizable theory where the Higgs mechanism preserves local gauge symmetry also in case of massive gauge bosons. The general proof of renormalizability ['t Hooft (1971)] and the impact on physics to elucidate the quantum structure of the electroweak interaction ['t Hooft and Veltman (1972)] were recognized by the Nobel prize in the year 1999. The virtue of renormalizability has motivated the calculation of the higher order contributions for electroweak observables, starting mid of the 1970 decade and continuing until today to reach the precision required at the LHC and possible future electron–positron colliders, like the International Linear Collider ILC [Moortgat-Pick (2015)] and the Future Circular Collider FCC [Abada (2019)].

That the loop contributions are sensitive to virtual heavy particles is known since the early work of Veltman [Veltman (1977)]. The loop diagrams contain the entire spectrum of the Standard Model, in particular also heavy particles like the top quark and the Higgs boson. Through their couplings to the gauge bosons they contribute to the W propagator

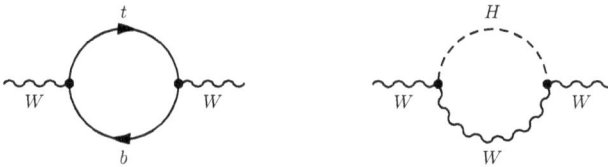

and similarly to the Z propagator, and thus they enter the S matrix elements with the propagators connecting two fermion lines. As a consequence, the observables calculated at one-loop order become dependent on the masses m_t and M_H also for processes at energies far below the scale for their direct production. Historically, before the discovery of top quark and Higgs boson, the precision measurements and comparison with the higher-order theory predictions provided an important indirect access to the mass range of those at that time unknown particles. Accordingly, the subsequent top discovery at the Tevatron [Abe (1995); Abachi (1995)] was a spectacular success of the electroweak theory, culminating in the Higgs boson discovery at the LHC [Aad (2012); Chatrchyan (2012b)]. The current experimental values for the masses are [Zyla (2020)]

$$m_t = 172.4 \pm 0.7\,\text{GeV}, \quad M_H = 125.10 \pm 0.14\,\text{GeV}. \tag{6.143}$$

By the discovery of the Higgs boson, the particle spectrum of the Standard Model has been completed. With the empirical knowledge of m_t and M_H, the imput quantities for calculating the precision observables are now fixed and the various observables can be uniquely calculated from the relations given above augmented by the higher-order contributions. The comparison of the predictions with the measured values hence have to be considered precision tests of the Standard Model.

6.5.4.1   *The vector boson mass correlation*

Exploiting the lowest-order relation (6.124) has shown that the obtained W mass and the M_W/M_Z mass ratio, respectively, are far away from the experimental results. Hence, the origin of that relation, the muon decay amplitude, has to be improved by loop contributions, which appear of the types indicated in Sec. 3.8: contributions to the gauge boson propagators (self-energies) and to the gauge boson–fermion vertices (vertex corrections), as well as exchange of two gauge bosons. As pointed out above, the decay amplitude becomes dependent on m_t and M_H, and the Fermi constant effectively summarizes a much more complicated expression than that in Eq. (6.122). As a consequence, the relation (6.124) has to be extended by a correction term

$$M_W^2 \left(1 - \frac{M_W^2}{M_Z^2} \right) = \frac{\pi \alpha}{\sqrt{2}\,G_F} \left[1 + \Delta r(m_t, M_H) \right] \tag{6.144}$$

where the quantity Δr summarizes the whole set of available loop corrections. The first calculations of the vector-boson masses [Antonelli (1980); Sirlin (1980); Veltman (1980)] were done at one-loop order. They are now complete at the two-loop level plus three- and four-loop leading terms, see [Awramik (2003)] for a summary in terms of a numerical parametrization and for more references. Choosing α, G_F, M_Z as input quantities, now augmented by the values of m_t, M_H as given above and α_s in Eq. (5.46), yields the prediction (the first error is the parametric uncertainty, the second error is the theoretical uncertainty from missing higher order contributions)

$$M_W = 80.361 \pm 0.005 \pm 0.004 \,\text{GeV}, \tag{6.145}$$

which is in good agreement with the experimental value in Eq. (6.121). By the accurate measurements of the vector boson masses the quantum correction Δr has been established with compelling significance,

$$\Delta r = 0.0367 \pm 0.0008, \tag{6.146}$$

in accordance with the theory prediction.

6.5.4.2 *Z resonance observables*

The set of precision observables at the Z resonance consists of two classes,

- the line shape observables: peak cross section, Z width, partial widths and ratios of partial widths;
- the various asymmetries, including τ polarization.

At lowest order, they can be expressed in terms of the fermion couplings to the Z boson, as outlined in Sec. 6.5.3. Higher-order electroweak contributions can be conveniently incorporated into *effective couplings* [Bardin (1995)] replacing the lowest-order couplings according to

$$v_f \to g_V^f(m_t, M_H), \quad a_f \to g_A^f(m_t, M_H) \tag{6.147}$$

in the respective formulae for the various observables. They include the loop contributions from self-energies and vertex corrections (two-boson exchange is negligible at the resonance) and thus they do depend also on m_t and M_H as extra input parameters. On top, one has to take into account additional *QED corrections* from photon radiation in the final states and *QCD corrections* from gluon radiation in case of hadronic final states (combined with virtual photon and virtual gluon contributions in either case), collected in multiplicative radiation factors $R_{V,A}^f$ for each decay channel. Accordingly, the partial widths (6.128) are modified as follows,

$$\Gamma_f = N_C^f \frac{\sqrt{2}\,G_F M_Z^3}{12\pi} \left((g_V^f)^2 \, R_V^f + (g_A^f)^2 \, R_A^f \right) \tag{6.148}$$

where the radiation factors are given as perturbative expansions in the electromagnetic and strong couplings α and $\alpha_s = \alpha_s(M_Z)$,

$$
R_{V,A}^f = 1 + \frac{3}{4}Q_f^2 \frac{\alpha}{\pi} + \cdots \qquad \text{for } f = \text{leptons}
$$

$$
R_{V,A}^f = 1 + \frac{3}{4}Q_f^2 \frac{\alpha}{\pi} + \frac{\alpha_s}{\pi} + \cdots \qquad \text{for } f = \text{quarks.}
$$

$$(6.149)$$

They are known up to $\mathcal{O}(\alpha_s^4)$ and $\mathcal{O}(\alpha^2)$. Moreover, finite mass terms have to be taken into account. Beyond one-loop order, the QCD radiation factors are different for the V and A parts of Γ_f [Chetyrkin (1996)].

A common way to write the effective couplings as close as possible to the lowest order expressions is to introduce correction factors ρ_f, κ_f in the following way,

$$
g_V^f = \sqrt{\rho_f}\,(I_3^f - 2\,\kappa_f s_W^2), \quad g_A^f = \sqrt{\rho_f}\,I_3^f \tag{6.150}
$$

with $\rho_f = \kappa_f = 1$ at lowest order. Instead of κ_f, an effective mixing angle can be introduced,

$$
\sin^2\theta_{\text{eff}}^f = \kappa_f s_W^2 = \kappa_f \left(1 - \frac{M_W^2}{M_Z^2}\right) \tag{6.151}
$$

which is specific for each type of fermions and which differs significantly from the lowest-order expression in terms of the W and Z mass ratio. A special role plays the leptonic mixing angle, $\sin^2\theta_{\text{eff}}^\ell$, obtained from measurements of all leptonic asymmetries as an average. It is one of the most precisely measured electroweak observables, with a very sensitive dependence on the Higgs boson mass on the theoretical side.

The Standard Model predictions of the Z observables are based on the complete two-loop terms and leading higher-order contributions. The current status is summarized in [Dubovyk (2019)] providing parametrizations for the various quantities in terms of the input parameters $\alpha, G_F, M_Z, m_t, M_H, \alpha_s$ and a comprehensive list of references. For comparison of theory and experiment, a selected set of Z resonance observables with experimental results [Schael (2005)] and theoretical predictions are put together in Table 6.4. The errors assigned to the predictions are the parametric uncertainty resulting from the input parameters, and the estimate of the theoretical uncertainty from missing higher order corrections (second error). Besides the line-shape observables, the averaged leptonic vector and axialvector couplings are listed in terms of ρ_ℓ and the effective mixing angle

Table 6.4 Z resonance observables (selection).

Observable	Measurement	Theory prediction
Γ_Z [GeV]	2.4952 ± 0.0023	$2.4945 \pm 0.0005 \pm 0.0004$
σ_0^{had} [nb]	41.540 ± 0.037	$41.492 \pm 0.005 \pm 0.006$
R_{had}	20.767 ± 0.025	$20.749 \pm 0.006 \pm 0.006$
R_b	0.21692 ± 0.00066	$0.21586 \pm 0.00003 \pm 0.0001$
ρ_ℓ	1.0050 ± 0.0010	$1.0053 \pm 0.0001 \pm 0.0002$
$\sin^2\theta_{\mathrm{eff}}^\ell$	0.23153 ± 0.00016	$0.23146 \pm 0.00003 \pm 0.00004$

$\sin^2\theta_{\mathrm{eff}}^\ell$, defined in Eqs. (6.150) and (6.151). From the measured values of $\sin^2\theta_{\mathrm{eff}}^\ell$ and M_W, M_Z one obtains

$$\kappa_\ell = 1.0382 \pm 0.0019.$$

The deviation from unity of both ρ_ℓ and κ_ℓ demonstrates the presence of quantum corrections with compelling significance, as Δr in Eq. (6.146) does.

The agreement between theoretical predictions and experimental results is impressive; this holds also for the full list given in [Schael (2005)]. Only a single quantity, the forward-backward asymmetry A_{FB}^b for b quarks, is more than two standard deviations off.

A global fit of all measured observables to the Standard Model predictions with M_H as the only free parameter results in a χ^2-distribution as displayed in Fig. 6.6 from which bounds on M_H can be read off:

$$M_H = 94_{-24}^{+29} \,\mathrm{GeV}, \quad M_H < 152 \,\mathrm{GeV} \text{ at } 95\% \text{ confidence level.}$$

The plot is from the website of the LEP Electroweak Working Group as an update of the one in the report [Schael (2005)], right before the discovery of a Higgs-like particle in summer 2012 by the LHC experiments ATLAS and CMS.

6.6 Higgs Bosons

The minimal model with a single scalar doublet is the simplest way to implement the electroweak symmetry breaking. The Higgs potential of the Standard Model, given in Eq. (6.78), involves two independent parameters μ and λ. Choosing the scalar doublet Φ in the unitary gauge (6.82),

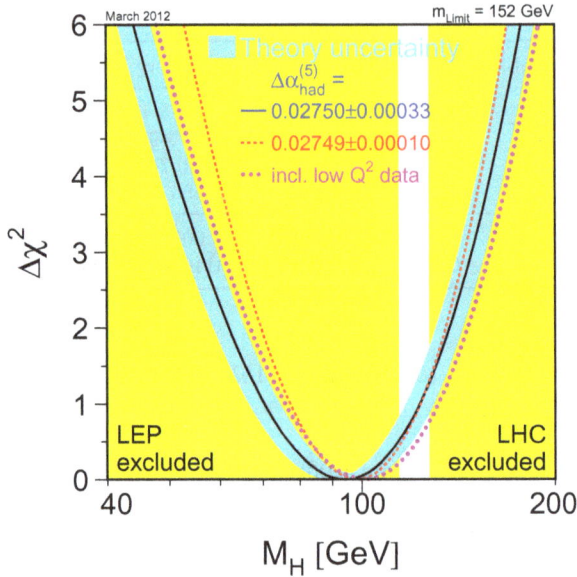

Fig. 6.6 Distribution of $\Delta\chi^2 = \chi^2(M_H) - \chi^2_{\min}$ from a global fit of precision observables.

the potential reads as follows (up to an irrelevant additive constant),

$$V = \left(-\mu^2 v + \frac{\lambda}{4} v^3\right) H + \frac{1}{2}\left(-\mu^2 + \frac{3}{4}\lambda v^2\right) H^2 + \frac{\lambda}{4} v\, H^3 + \frac{\lambda}{16}\, H^4. \tag{6.152}$$

With the minimum condition (6.79) for v, the term linear in H vanishes and one obtains

$$V = \frac{1}{2}\left(2\mu^2\right) H^2 + \frac{\mu^2}{v}\, H^3 + \frac{\mu^2}{4v^2}\, H^4. \tag{6.153}$$

The real scalar field H describes neutral particles of spin 0, the Higgs bosons, with mass M_H determined by the coefficient of the quadratic term,

$$M_H = \mu\sqrt{2}. \tag{6.154}$$

Hence, the potential can be written as follows

$$V = \frac{M_H^2}{2}\, H^2 + \frac{M_H^2}{2v}\, H^3 + \frac{M_H^2}{8v^2}\, H^4 \tag{6.155}$$

in terms of M_H and v. The vacuum expectation value v is determined by the gauge sector, in particular by the Fermi constant, as shown in Eq. (6.123). M_H is independent and cannot be predicted but has to be taken from experiment. Thus, in the Standard Model the mass M_H of the Higgs boson appears as a free parameter, sufficient to describe the dynamics of the scalar sector. The part of the electroweak Lagrangian involving the Higgs field is given by $\mathcal{L}_H$ in Eq. (6.80) and by $\mathcal{L}_Y$ from Eqs. (6.92) and (6.100), and can be cast into the following form,

$$\mathcal{L}_H + \mathcal{L}_Y = \frac{1}{2}(\partial_\mu H)(\partial^\mu H) - \frac{M_H^2}{2} H^2 - \frac{M_H^2}{2v} H^3 - \frac{M_H^2}{8v^2} H^4$$

$$+ \left(M_W^2\, W_\mu^+ W^{-\mu} + \frac{M_Z^2}{2} Z_\mu Z^\mu \right) \left(1 + \frac{H}{v} \right)^2$$

$$- \sum_f m_f \overline{\psi}_f \psi_f \left(1 + \frac{H}{v} \right) \tag{6.156}$$

involving the interactions of the Higgs field with massive fermions and gauge bosons, as well as the Higgs self interaction. The scalar self-interaction is new, the others have already been discussed in the previous sections. The cubic and quartic Higgs interactions augment the list of Feynman rules by scalar vertices of the triple and quartic type. They have already been derived for generic coupling constants κ and λ in Eq. (4.142) and can be taken over from there with the specification

$$\kappa = -\frac{M_H^2}{2v}, \quad \lambda = -\frac{M_H^2}{8v^2}.$$

With this final step, all the interaction vertices of the electroweak Standard Model are determined, in particular those relevant for the production and decay processes of Higgs bosons (see e.g. [Djouadi (2008)] for a review). They provide the basis for the searches that were done at LEP and the Tevatron yielding exclusion limits, as well as for the successful search at the LHC and for ongoing and forthcoming precision studies.

6.6.1 *Higgs boson production*

Higgs bosons can be produced in proton–proton collisions through various mechanisms at the partonic level. The main partonic processes for Higgs boson production are

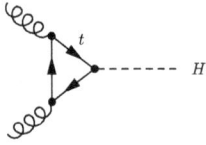

gluon fusion

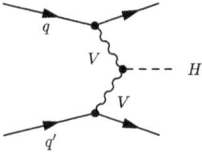

vector boson fusion $(VV = ZZ, W^+W^-)$

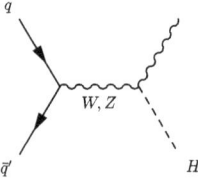

associated Higgs–V production

and the corresponding hadronic cross sections follow according to the rules given in Sec. 5.2 invoking the gluon and quark distributions. Gluon fusion through the top-quark loop yields the largest cross section at the LHC because of the dominating gluon distribution (see Fig. 5.6).

Note that the Feynman diagrams above display only the lowest order matrix elements. For reliable predictions higher order QCD contributions have been calculated, up to 4-loop order (N3LO), as well as electroweak contributions of the next order (NLO). Both are mandatory to reach the precision for the correct interpretation of the observed experimental signatures. Figure 6.7 shows the calculated cross sections for the various parton processes. It contains also the $t\bar{t}H$ and $b\bar{b}H$ final states, which are important for probing the heavy quark Yukawa couplings directly since H is radiated from the top and bottom quarks, respectively.

6.6.2 *Higgs boson decays*

Since the Higgs boson is an unstable particle, the experimental signal is determined by the product

$$\sigma(AB \to H) \cdot BR(H \to X) \qquad (6.157)$$

of the production cross section $\sigma(AB \to H)$ from initial-state partons A, B and the branching ratio for the decay of the Higgs boson into a specific final

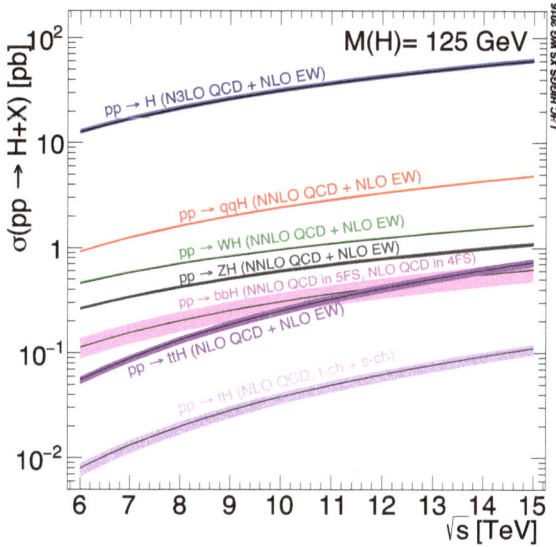

Fig. 6.7 Cross section for Higgs boson production at the LHC corresponding to various parton processes, versus the collider energy $\sqrt{S}$ [de Florian (2017)].

state X, defined by

$$BR(H \to X) = \frac{\Gamma(H \to X)}{\Gamma_H}$$

with the total width Γ_H. Therefore, the various decay modes have to be taken into account as well. By the Yukawa and Higgs–gauge interactions, Higgs bosons decay directly into fermion pairs, $H \to f\bar{f}$, and into gauge bosons, $H \to WW, ZZ$, described by tree-like amplitudes

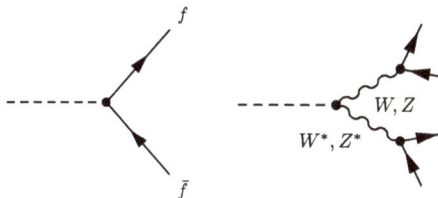

at lowest order. Since M_H is well below the WW threshold, Higgs bosons decay predominantly into $b\bar{b}$ quarks, owing to the largest Yukawa coupling in the kinematically allowed fermionic decay channels. This signal, however, is experimentally difficult to identify because it is covered by a huge background of QCD-generated b-quarks. The decay modes $H \to WW, ZZ$

with subsequent decays into 4 fermions, especially leptons, make detection easier; in particular, $H \to ZZ \to 4\,\text{leptons}$ has been one of the discovery channels. For kinematical reasons, one of the vector bosons has to be off-shell, indicated by a V^* in the diagram.

The other decay process with a clear signal for experimental studies is the loop-induced decay into two photons, $H \to \gamma\gamma$,

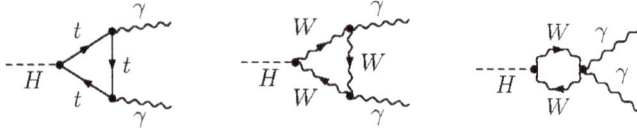

which in lowest order is described by one-loop diagrams with virtual top quarks and W bosons. The decay rate carries the loop factor $\frac{\alpha}{\pi}$ and is accordingly suppressed by $\mathcal{O}(10^{-3})$ relative to the tree-level decays (see Fig. 6.8 for the branching ratios of the various decay modes). In spite of being a rare decay, the $\gamma\gamma$ channel has been the other discovery channel because of its clear signature. The high Higgs boson production rate at the LHC ensures that a sufficiently large number of events is available in the detectors, the analysis, however, requires an enormous effort. Also from

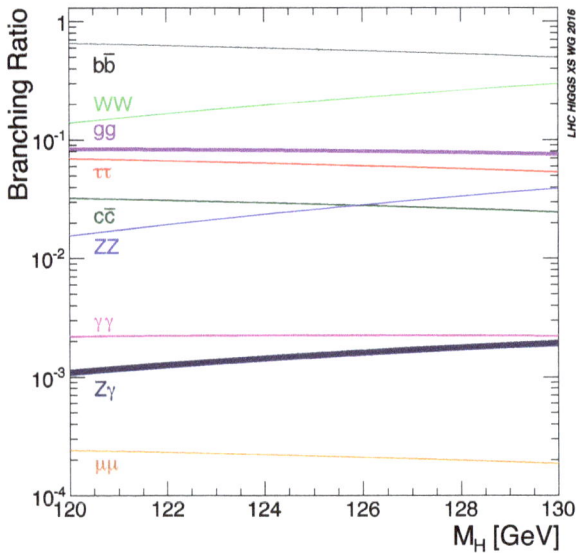

Fig. 6.8   Branching ratios for the various decay modes of the Higgs bosons [de Florian (2017)].

the theoretical side, a substantial amount of work has been invested to get the decay rates reliably predicted by including the necessary higher order QCD and electroweak contributions. Summing up all partial decay withs yields the total width of $\Gamma_H = 4.1\,\text{MeV}$ for a Higgs boson at $125\,\text{GeV}$, appearing as a narrow resonance.

6.6.3 *Experimental studies*

At the LHC, data were collected during Run 1 at $\sqrt{S} = 7$ and $8\,\text{TeV}$ from 2009 to 2012 and during Run 2 at $\sqrt{S} = 13\,\text{TeV}$ from 2015 to 2018. In the year 2012 a scalar particle was discovered by the experiments ATLAS and CMS via the decay modes into $\gamma\gamma$ and $ZZ \to 4$ leptons showing the properties of the Standard Model Higgs boson. Subsequently, from measurements of the invariant mass distributions $m_{\gamma\gamma}$ and $m_{4\text{lept}}$ the mass has been determined and has meanwhile reached an impressive accuracy [Zyla (2020)]

$$M_H = 125.10 \pm 0.14\,\text{GeV}. \qquad (6.158)$$

As an example, Fig. 6.9 displays the $m_{4\text{lept}}$ distribution from data collected during Run 2 by ATLAS.

Most of the other production and decay channels shown in Figs. 6.7 and 6.8 have also been observed; by now the H couplings to gauge bosons and

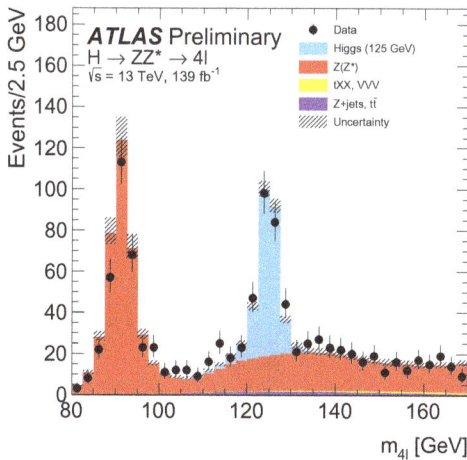

Fig. 6.9 Invariant mass distributions of the final fermions in $H \to ZZ^* \to 4f$ [ATLAS (2020)]. The Higgs boson corresponds to the blue region around $125\,\text{GeV}$.

to the fermions of the third generation b, τ are established with an accuracy of the order of 10–20%. For quantitative tests the following ratio, called the *signal strength*, has been used for each mode with H decaying to X,

$$\mu_X = \frac{\sigma(PP \to H) \cdot BR(H \to X)}{\left[\sigma(PP \to H) \cdot BR(H \to X)\right]_{\text{SM}}} \qquad (6.159)$$

which is the ratio of the measured rate and the corresponding Standard Model prediction. This ratio has to be unity for the case that the new boson is the Higgs particle of the Standard Model. All the measurements so far are compatible with $\mu_X = 1$, with an accuracy of $\mathcal{O}(10)\%$ for the best accessible final states $\gamma\gamma$ and ZZ. The individual signal strengths for the various final states that have been observed are put together in Table 6.5. The values are averages of ATLAS and CMS results, given in the Particle Data Group Review 2020 [Zyla (2020)]. The current data are not yet sufficient to determine the top–Higgs Yukawa coupling; just an upper limit $\left|g_t/g_t^{\text{SM}}\right| < 1.7$ can be assigned.

The global signal strength obtained as an inclusive quantity from all modes has been determined by the two experiments [Aad (2020); Sirunyan (2019)] as follows,

$$\mu = 1.11^{+0.09}_{-0.08} \text{ (ATLAS)}, \quad \mu = 1.17 \pm 0.10 \text{ (CMS)}.$$

The total width can be measured only indirectly, through interference between the resonance (signal) and non-resonance amplitudes. Accordingly, the experimental constraints are not very stringent, $\Gamma_H < 13\,\text{MeV}$ at the 95% confidence level, showing consistency with the Standard Model expectation of $4\,\text{MeV}$.

At the coming LHC runs the uncertainty of the measured Higgs observables will be further reduced, both by experimental and theoretical progress.

Table 6.5 Signal strengths for different final states [Zyla (2020)].

Final state	Signal strength
ZZ^*	$1.20^{+0.12}_{-0.11}$
WW^*	1.08 ± 0.12
$\gamma\gamma$	$1.11^{+0.10}_{-0.09}$
$b\bar{b}$	1.04 ± 0.13
$\tau^+\tau^-$	$1.15^{+0.16}_{-0.15}$
$\mu^+\mu^-$	0.6 ± 0.8
$t\bar{t}H \;\; (H \to \gamma\gamma, WW, ZZ, b\bar{b})$	1.28 ± 0.20

The phase of the LHC with high luminosity, the HL-LHC, expected to go into operation in 2026, will improve the accuracy by a factor of 5–10 with respect to the present data. Another future task is to establish the Higgs-boson self coupling, an important quantity that determines the Higgs potential. It can be measured in Higgs-boson pair production, a process which has already been observed, but needs significantly more data to pin down the self coupling. One should not forget that the mechanism of the Standard Model is only one option for electroweak symmetry breaking, and also other scenarios with more than one scalar boson might be realized in nature. Since the Higgs sector is currently the least understood part of the Standard Model, further investigations may reveal a more complex structure than the minimal version with a single Higgs particle. Besides the LHC program, there are also other concepts of prospective accelerators based on electron–positron collisions at energies high enough to study Higgs bosons and their properties with high precision [Aihara (2019)]. The aim is to either confirm the scalar particle at 125 GeV as part of the Standard Model or to find significant deviations from the predictions, discoveries included, making a revision of the Standard Model necessary.

6.7 Open Questions

The Standard Model has proven very successful in describing a large variety of phenomena from low to high energies at an impressive level of accuracy on both the theoretical and the experimental side. In spite of this success, there is a list of shortcomings that motivate the quest for physics beyond the Standard Model.

A rather direct augmentation is enforced by the need for accommodating massive neutrinos. The Standard Model in its strictly minimal version is incomplete with respect to a mass term for neutrinos. Neutrino mass terms can be added on top of the current formulation without touching the basic architecture of the Standard Model. Besides this rather immediate modification one is confronted, however, with a series of basic conceptual problems,

- the *hierarchy problem*, i.e. the enormous hierarchy $M_{\text{Pl}} \gg v$ between the Planck scale $M_{\text{Pl}} \sim 1/\sqrt{G_N}$ (set by Newton's constant G_N) and the electroweak scale $v \sim 1/\sqrt{G_F}$, and the smallness of the Higgs boson mass of $\mathcal{O}(v)$, which is not protected against large quantum corrections of $\mathcal{O}(M_{\text{Pl}})$;

- the large number of free parameters (gauge couplings, vacuum expectation value, Higgs boson mass, fermion masses, CKM matrix elements), which are not predicted but have to be taken from experiments;
- the pattern that occurs in the arrangement of the fermion masses and the texture of fermion mixing;
- the quantization of the electric charge, or the values of the hypercharge, respectively;
- the non-unification of the running strong, electromagnetic and weak coupling constants at some high-energy scale, which is against the spirit of a further unification of the fundamental forces;
- the missing way to connect to gravity.

Moreover, there are also phenomenological shortcomings, like missing answers to the questions for

- the nature of dark matter that constitutes the largest fraction of matter in the universe,
- the origin of the baryon asymmetry of the universe, for which the amount of CP violation in the Standard Model is not sufficient.

A crucial aspect is the still unresolved dynamics of electroweak symmetry breaking, leaving many more options conceivable than the minimal solution within the Standard Model. In view of missing direct experimental signals for physics beyond the Standard Model so far, quite a variety of scenarios for modifications or extensions are being considered, such as those assuming either a further substructure (composite models) or an additional symmetry, like supersymmetry. Further input is needed from the experiments in the forthcoming runs at the LHC.

Bibliography

Aad, G. *et al.* [ATLAS] (2012). Observation of a new particle in the search for the Standard Model Higgs boson with the ATLAS detector at the LHC, *Phys. Lett. B* **716**, 1 [arXiv:1207.7214 [hep-ex]].

Aad, G. *et al.* [ATLAS] (2020). Combined measurements of Higgs boson production and decay using up to 80 fb^{-1} of proton-proton collision data at $\sqrt{s} = 13$ TeV collected with the ATLAS experiment, *Phys. Rev. D* **101**, 012002 [arXiv:1909.02845 [hep-ex]].

Abachi, S. *et al.* [D0] (1995). Observation of the top quark, *Phys. Rev. Lett.* **74**, 2632 [arXiv:hep-ex/9503003 [hep-ex]].

Abada, A. *et al.* [FCC] (2019). FCC-ee: The Lepton Collider: Future Circular Collider Conceptual Design Report Volume 2, *Eur. Phys. J. ST* **228**, 261.

Abe, K. *et al.* [SLD] (1994). Precise measurement of the left-right cross-section asymmetry in Z boson production by e^+e^- collisions, *Phys. Rev. Lett.* **73**, 25 [arXiv:hep-ex/9404001 [hep-ex]].

Abe, K. *et al.* [CDF] (1995). Observation of top quark production in $\bar{p}p$ collisions, *Phys. Rev. Lett.* **74**, 2626 [arXiv:hep-ex/9503002 [hep-ex]].

Abi, B. *et al.* [Muon g-2] (2021). Measurement of the positive muon anomalous magnetic moment to 0.46 ppm, *Phys. Rev. Lett.* **126**, 141801 [arXiv:2104.03281 [hep-ex]].

Aihara, H. *et al.* [ILC] (2019). The International Linear Collider. A Global Project, arXiv:1901.09829 [hep-ex].

Allaby, J. V. *et al.* [CHARM] (1987). A precise determination of the electroweak mixing angle from semi-leptonic neutrino scattering, *Phys. Lett. B* **177**, 446.

Altarelli, G and Parisi, G. (1977). Asymptotic freedom in parton language, *Nucl. Phys. B* **126**, 298.

Antonelli, F., Consoli, M. and Corbo, G. (1980). One loop correction to vector boson masses in the Glashow-Weinberg-Salam model of electromagnetic and weak interactions, *Phys. Lett. B* **91**, 90.

Aoyama, T., Kinoshita, T. and Nio, M. (2019). Theory of the anomalous magnetic moment of the electron, *Atoms* **7**, 28.

Aoyama, T. *et al.* (2020). The anomalous magnetic moment of the muon in the Standard Model, *Phys. Rept.* **887**, 1 [arXiv:2006.04822 [hep-ph]].

Arnison, G. *et al.* [UA1] (1983). Experimental observation of isolated large transverse energy electrons with associated missing energy at $\sqrt{s} = 540$ GeV, *Phys. Lett. B* **122**, 10. Experimental observation of lepton pairs of invariant mass around 95 GeV/c^2 at the CERN SPS collider, *Phys. Lett. B* **126**, 398.

ATLAS (2020). Measurement of the Higgs boson mass in the $H \to ZZ^* \to 4\ell$ decay channel with $\sqrt{s} = 13$ TeV pp collisions using the ATLAS detector at the LHC, ATLAS-CONF-2020-005.

Awramik, M., Czakon, M., Freitas, A. and Weiglein G. (2003). Precise prediction for the W boson mass in the standard model, *Phys. Rev. D* **69**, 053006 [arXiv:hep-ph/0311148 [hep-ph]].

Banner, M. *et al.* [UA2] (1983). Observation of single isolated electrons of high transverse momentum in events with missing transverse energy at the CERN $\bar{p}p$ collider, *Phys. Lett. B* **122** (1983), 476. Evidence for $Z^0 \to e^+ e^-$ at the CERN $\bar{p}p$ collider, *Phys. Lett. B* **129** (1983), 130.

Bardin, D., Hollik, W. and Passarino, G. (eds.) (1995). Reports of the Working Groups on Precision Calculations for the Z Resonance, *CERN Yellow Reports* **95-03** [arXiv:hep-ph/9709229 [hep-ph]].

Becchi, C., Rouet, A. and Stora, R. (1976). Renormalization of gauge theories, *Annals Phys.* **98**, 287.

Bennett, C. W. *et al.* [Muon g-2] (2006). Final report of the muon E821 anomalous magnetic moment measurement at BNL, *Phys. Rev. D* **73**, 072003 [arXiv:hep-ex/0602035 [hep-ex]].

Bethe, H. A. (1947). The electromagnetic shift of energy levels, *Phys. Rev.* **72**, 339.

Blondel, A. *et al.* [CDHS] (1990). Electroweak parameters from a high statistics neutrino nucleon scattering experiment, *Z. Phys. C* **45**, 361.

Breidenbach, M. *et al.* (1969). Observed behavior of highly inelastic electron-proton scattering, *Phys. Rev. Lett.* **23**, 935.

Cabibbo, N. (1963). Unitary symmetry and leptonic decays, *Phys. Rev. Lett.* **10**, 531.

Chatrchyan, S. *et al.* [CMS] (2012a). Performance of CMS muon reconstruction in pp collision events at $\sqrt{s} = 7$ TeV, *JINST* **7**, P10002 [arXiv:1206.4071 [physics.ins-det]].

Chatrchyan, S. *et al.* [CMS] (2012b). Observation of a new boson at a mass of 125 GeV with the CMS experiment at the LHC, *Phys. Lett. B* **716**, 30 [arXiv:1207.7235 [hep-ex]].

Chetyrkin, K. G., Kühn, J. H. and Kwiatkowski, A. (1996). QCD corrections to the $e^+ e^-$ cross section and the Z boson decay rate, *Phys. Rept.* **277**, 189 [arXiv:hep-ph/9503396 [hep-ph]].

de Florian, D. *et al.* [LHC Higgs Cross Section Working Group] (2017). Handbook of LHC Higgs Cross Sections: 4. Deciphering the nature of the Higgs sector, *CERN Yellow Reports: Monographs, Vol. 2/2017*, CERN-2017-002-M (CERN, Geneva, 2017), https://doi.org/10.23731/CYRM-2017-002 [arXiv:1610.07922 [hep-ph]].

Denner, A. (1993). Techniques for calculation of electroweak radiative corrections at the one loop level and results for W physics at LEP-200, *Fortsch. Phys.* **41**, 307 [arXiv:0709.1075 [hep-ph]].

Djouadi, A. (2008). The anatomy of electroweak symmetry breaking. I: The Higgs boson in the standard model, *Phys. Rept.* **457**, 1 [arXiv:hep-ph/0503172 [hep-ph]].

Dokshitzer, Y. L. (1977). Calculation of the structure functions for deep inelastic scattering and e+ e- annihilation by perturbation theory in Quantum Chromodynamics, *Sov. Phys. JETP* **46**, 641.

Dubovyk I., Freitas, A., Gluza J., Riemann, T. and Usovitsch J. (2019). Electroweak pseudo-observables and Z-boson form factors at two-loop accuracy, *JHEP* **08**, 113 [arXiv:1906.08815 [hep-ph]].

Englert, F. and Brout, R. (1964). Broken symmetry and the mass of gauge vector mesons, *Phys. Rev. Lett.* **13**, 321.

Faddeev, L. D. and Popov, V. N. (1967). Feynman diagrams for the Yang-Mills field, *Phys. Lett. B* **25**, 29.

Feynman, R. P. (1969). Very high-energy collisions of hadrons, *Phys. Rev. Lett.* **23**, 1415.

Fritzsch, H. and Gell-Mann, M. (1972). Current algebra: Quarks and what else?, in *Proceedings of the XVI International Conference on High Energy Physics*, Vol. 2, 164 [arXiv:hep-ph/0208010 [hep-ph]].

Gell-Mann, M. (1962). Symmetries of baryons and mesons, *Phys. Rev.* **125**, 1067.

Gell-Mann, M. (1964). A schematic model of baryons and mesons, *Phys. Lett.* **8**, 214.

Glashow, S. L. (1961). Partial symmetries of weak interactions, *Nucl. Phys.* **22**, 579.

Glashow, S. L., Iliopoulos, J. and Maiani, L. (1970). Weak interactions with lepton-hadron symmetry, *Phys. Rev. D* **2**, 1285.

Goldstone, J. (1961). Field theories with superconductor solutions, *Nuovo Cim.* **19**, 154.

Goldstone, J., Salam, A. and Weinberg, S. (1962). Broken Symmetries, *Phys. Rev.* **127**, 965.

Grange, J. *et al.* [Muon g-2] (2015). Muon (g-2) Technical Design Report [arXiv:1501.06858 [physics.ins-det]].

Gribov, V. N. and Lipatov, L. N. (1972). Deep inelastic e p scattering in perturbation theory, *Sov. J. Nucl. Phys.* **15**, 438.

Gross, D. J. and Wilczek, F. (1973). Ultraviolet behavior of nonabelian gauge theories, *Phys. Rev. Lett.* **30**, 1343.

Guralnik, G. S., Hagen, C. R. and Kibble, T. W. B. (1964). Global conservation laws and massless particles, *Phys. Rev. Lett.* **13**, 585.

Hanneke, D., Fogwell, S. and Gabrielse, G.(2008). New measurement of the electron magnetic moment and the fine structure constant, *Phys. Rev. Lett.* **100**, 120801 [arXiv:0801.1134 [physics.atom-ph]].

Hasert, F. J. *et al.* [GARGAMELLE] (1973). Search for elastic ν_μ electron scattering, *Phys. Lett. B* **46**, 121; Observation of neutrino like interactions without muon or electron in the Gargamelle neutrino experiment, *Phys. Lett. B* **46**, 138.

Higgs, P. W. (1964). Broken symmetries and the masses of gauge bosons, *Phys. Rev. Lett.* **13**, 508.

Jegerlehner, F. (2019). Variations on photon vacuum polarization, *EPJ Web Conf.* **218**, 01003 [arXiv:1711.06089 [hep-ph]].

Jegerlehner, F. (2017). The anomalous magnetic moment of the muon (2nd edition), *Springer Tracts in Modern Physics*, **274**.

Karplus, R., Klein, R. and Schwinger, J. (1952). Electrodynamic displacement of atomic energy levels. 2. Lamb Shift, *Phys. Rev.* **86**, 288.

Kobayashi, M. and Maskawa, T. (1973). CP violation in the renormalizable theory of weak interaction, *Prog. Theor. Phys.* **49**, 652.

Lamb, W. E. and Retherford, R. C. (1947). Fine structure of the hydrogen atom by a microwave method, *Phys. Rev.* **72**, 241.

McFarland, K. S. *et al.* [CCFR] (1998). A precision measurement of electroweak parameters in neutrino-nucleon scattering, *Eur. Phys. J. C* **1**, 509.

Moortgat-Pick, G. *et al.* (2015). Physics at the e+ e- Linear Collider, *Eur. Phys. J. C* **75**, 371 [arXiv:1504.01726 [hep-ph]].

Parker, R. H., Yu, C., Zhong, W., Estey, B. and Müller, H. (2018). Measurement of the fine-structure constant as a test of the Standard Model, *Science* **360**, 191 [arXiv:1812.04130 [physics.atom-ph]].

Peskin, M. E. and Schroeder, D. V. (1995). An Introduction to Quantum Field Theory.

Politzer, H. D. (1973). Reliable perturbative results for strong interactions?, *Phys. Rev. Lett.* **30**, 1346.

Salam, A. (1968). Weak and electromagnetic interactions, Contribution to: 8th Nobel Symposium, *Conf. Proc. C* **680519**, 367.

Schael, S. *et al.* [ALEPH, DELPHI, L3, OPAL, SLD, LEP Electroweak Working Group, SLD Electroweak Group and SLD Heavy Flavour Group] (2006). Precision electroweak measurements on the *Z* resonance, *Phys. Rept.* **427**, 257 [arXiv:hep-ex/0509008 [hep-ex]], http://lepewwg.web.cern.ch/LEPEWWG.

Schael, S. *et al.* [ALEPH, DELPHI, L3, OPAL and LEP Electroweak] (2013). Electroweak measurements in electron-positron collisions at W-boson-pair energies at LEP, *Phys. Rept.* **532**, 119 [arXiv:1302.3415 [hep-ex]].

Schwinger, J. S. (1948). On quantum electrodynamics and the magnetic moment of the electron, *Phys. Rev.* **73**, 416.

Sirlin, A. (1980). Radiative corrections in the $SU(2)_L \times U(1)$ theory: a simple renormalization framework, *Phys. Rev. D* **22**, 971.

Sirunyan, A. M. *et al.* [CMS] (2019). Combined measurements of Higgs boson couplings in proton–proton collisions at $\sqrt{s} = 13$ TeV, *Eur. Phys J. C* **79**, 421 [arXiv:1809.10733 [hep-ex]].

Tanabashi, M. *et al.* [Particle Data Group] (2018). Review of Particle Physics, *Phys. Rev. D* **98**, 030001. https://pdg.lbl.gov/2018/.

't Hooft, G. (1971). Renormalizable Lagrangians for massive Yang-Mills fields, *Nucl. Phys. B* **35**, 167.

't Hooft, G. and Veltman M. J. G. (1972). Regularization and renormalization of gauge fields, *Nucl. Phys. B* **44**, 189.

Tyutin, I. V. (1975). Gauge invariance in field theory and statistical physics in operator formalism, *Lebedev Physics Institute preprint* 39 (1975) [arXiv:0812.0580 [hep-th]].

Veltman M. J. G. (1977). Limit on mass differences in the Weinberg model, *Nucl. Phys. B* **123**, 89.

Veltman M. J. G. (1980). Radiative corrections to vector boson masses, *Phys. Lett. B* **91**, 95.

Vilain, P. *et al.* [CHARM-II] (1994). Precision measurement of electroweak parameters from the scattering of muon-neutrinos on electrons, *Phys. Lett. B* **335**, 246.

Weinberg, S. (1967) A model of leptons, *Phys. Rev. Lett.* **19**, 1264.

Zeller, G. P. *et al.* [NuTeV] (2002). A precise determination of electroweak parameters in neutrino nucleon scattering, *Phys. Rev. Lett.* **88**, 091802 [erratum: *Phys. Rev. Lett.* **90**, 239902] [arXiv:hep-ex/0110059].

Zyla, P. A. *et al.* [Particle Data Group] (2020). Review of Particle Physics, *Prog. Theor. Exp. Phys.* **2020**, 083C01. https://pdg.lbl.gov.

Index